TECHNOLOGY AND SOCIETY

TECHNOLOGY AND SOCIETY

Advisory Editor
DANIEL J. BOORSTIN, author of
The Americans and Director of
The National Museum of History
and Technology, Smithsonian Institution

WILLIS HAVILAND CARRIER

Father of
Air Conditioning

by Margaret Ingels

ARNO PRESS
A NEW YORK TIMES COMPANY
New York • 1972

Reprint Edition 1972 by Arno Press Inc.

Copyright © 1952 Doubleday & Company, Inc.
Reprinted by permission of Carrier Corporation

Reprinted from a copy in The University of
Illinois Library

Technology and Society
ISBN for complete set: 0-405-04680-4
See last pages of this volume for titles.

Manufactured in the United States of America

Library of Congress Cataloging in Publication Data

Ingels, Margaret.
    Willis Haviland Carrier, father of air conditioning.

    (Technology and society)
    Bibliography: p.
    1. Carrier, Willis Haviland, 1876-1950. 2. Air
conditioning--History. I. Series.
TH7687.I54 1972   697.9'3'0924[B]   72-5056
ISBN 0-405-04708-8

WILLIS HAVILAND CARRIER

Father of
Air Conditioning

by Margaret Ingels

1952 • COUNTRY LIFE PRESS • GARDEN CITY

Designed and manufactured at Country Life Press, Garden City

Copyright 1952 Doubleday & Company, Inc.

FOREWORD

By Cloud Wampler

My association with Willis Carrier, the company which bears his name, and the industry he founded had a strange beginning in the early thirties.

I was a banker in Chicago and our firm managed the building in which Carrier Corporation had its middle western offices. So the company, then struggling under the combined load of the Great Depression and a merger, came to us requesting a reduction in rent.

It soon became evident to our real estate people that rent in Chicago was only part of a larger and more pressing problem for Carrier Corporation. The immediate task was to cut *all* expenses —to make ends meet. And because of my industrial experience, I quickly found myself in the middle of things.

From that time on, destiny seems to have done more than its usual share in bringing Carrier and me closer together. But the basic fact is that the infant air conditioning industry intrigued me no end, and I was sure that it had tremendous potentials. Perhaps because of this, President Irvine Lyle asked me to become financial adviser to the Corporation. I accepted—and the stage was set for my unforgettable first meeting with "The Chief," who was then Chairman of the Board.

I had already been told that Dr. Carrier was a genius and that his talents lay in the field of science and invention rather than in operation and finance. All the same, I wasn't prepared for what happened when we sat down to discuss the company's problems.

Right off the bat, Dr. Carrier made it perfectly clear that he held a dim view of bankers. I couldn't possibly know much about the air conditioning industry—he told me frankly—and what there was to know would take me many years to learn. His attitude was more significant than I then realized: I was not an engineer.

In spite of all this, I felt myself tremendously drawn to the man. For he certainly had a way with him. Among other things, his physical appearance was striking. Almost six feet in height, he had powerful shoulders, a majestic head topped with tousled hair, the weathered complexion of the outdoor man he was, and a face filled with lines of character.

In his manner of speaking Dr. Carrier was equally compelling. I remember so well the ring in his voice when he said to me that day: "We will not do less research and development work"; "We will not discharge the people we have trained"; and "We will all work for nothing if we have to."

This last remark was especially characteristic. But it was also an indication of how his people felt toward "The Chief" and of his faith in them.

Without question, the men and women of Carrier Corporation loved Willis Carrier. But there was something beyond affection, and that was great respect. This found expression in many ways; for example, the name by which he was always known. As one of his early associates explained: "He was *The Chief* because we accepted him as our leader both at work and at play."

How well that name fitted the man! For he was absolutely fearless—physically, mentally, and spiritually. And this quality had a great bearing upon Dr. Carrier's success in the engineering field. He did not hesitate to move into areas that were uncharted. He was always pushing across new frontiers. His entire life was that of the pioneer. No wonder he became the founder of a new industry.

But there are other reasons why Dr. Carrier was so very special. First, he was one hundred per cent honest intellectually. Second, he possessed great knowledge, supplemented by a wonderful intuitive sense. Third, he always went directly to the heart of a problem—often with an unconventional approach. Fourth, he had an uncanny faculty for selecting a project that promised the fulfillment of a human need and was, therefore, commercially attractive. He sensed demand and then developed something to meet it.

It was in 1915 that Dr. Carrier and Irvine Lyle, together with five other young engineers, formed Carrier Engineering Corporation. At that time the first World War was under way and I think it safe to say that fear was everywhere. But this bunch of youngsters, scraping together everything they could—$32,600—embarked with high enthusiasm upon a new and seemingly risky venture. And what they accomplished is one of the finest examples of the workings of the American free-enterprise system.

Here is the place, I believe, to pay tribute to a great team—led by Carrier and Lyle. Carrier, the scientist and inventor; Lyle, the business manager and salesman. And the other five—Murphy, Lewis, Stacey, Heckel, and the younger Lyle. Each of these played important but different parts. But the teamwork was good, for there was a good captain, a good quarterback, and good players. And before long, there were more good players. The "original seven" taught others and this was especially true of Dr. Carrier who, until the very month of his death, was engaged in building a great engineering organization.

Bringing others along—that is where Willis Carrier was at his best, although it may not have appeared so. For his teaching followed no formal pattern. It took place from day to day as those with problems came to him throughout his long productive years. He always "had the time." He was constantly interested in developing young people, and he never failed to inspire them.

Great as were his scientific achievements—and they were great indeed—it was Dr. Carrier's ability to teach and inspire that created the finest legacy to the company bearing his name, which I now have the honor to head.

If Willis Carrier were with us today, I believe he would ask that this book be dedicated in a certain way. Out of our years of close association, I shall try to give it as he would:

> *To the young engineers who are*
> *today creating a better tomorrow.*

WILLIS Haviland Carrier, the father of air conditioning, was born on a farm near Angola, in the western part of New York State, on November 26, 1876. He once described his forebears as "a rugged and adventurous people, with courage to try the unknown"—qualities which he himself possessed in no small measure.

The first Carrier in America was Thomas, who arrived in Massachusetts around 1663. There is historical evidence that he was born in Wales in 1622 and that he was a political refugee who assumed the name "Carrier" on coming to America. One story is that he was Richard Whaley, a political leader when Charles I of England was beheaded. Another holds that his last name was Morgan and that he was one of the king's bodyguards, seven feet four inches tall.

New England town records embody traditions handed down about him: that he was fabulously fleet of foot even after he was more than one hundred years old; that he would carry corn on his shoulder eighteen miles to a mill, walking very fast and stopping only once en route; and that up to his death, which came when he was 113 years old, he still walked erect and was neither gray nor bald.

Thomas Carrier married Martha Allen, who was his equal in all except longevity. She was a daughter of Andrew Allen, a first settler of Andover, Massachusetts. Abbot's *History of Andover* describes her as "plain and outspoken in speech, of remarkable strength of mind, a keen sense of justice, and a sharp tongue." After standing up against the Andover town fathers in a boundary dispute, she was accused of being a witch. Two of her sons, aged thirteen and ten, were hung by their heels until they too testified against her. Cotton Mather denounced her as a "rampant hag" whom the Devil had promised "should be the queen of Hell." She was arrested, convicted and, on August 19, 1692, hanged on Salem's Gallows Hill. Later it was recorded that of all the New Englanders charged with witchcraft, "Martha Carrier was the only one, male or female, who did not at some time or other make an admission or confession."

The Carriers lived in New England until 1799 when Willis Carrier's great-grandparents, then bride and groom, joined an ox-team train of settlers pushing west through the Mohawk Valley. They settled in Madison County, New York, and then in 1836 moved on west again to Erie County. There they purchased the farm that became the birthplace and childhood home of Willis Carrier. His father was Duane Carrier, who taught music to the Indians, tried running a general store, was for a short time a postmaster, then settled down to farming and married Elizabeth Haviland. Her forefathers had settled in New England in the seventeenth century and she was a "birthright" Quaker— the first in her family to marry outside her faith.

Willis Carrier grew up as the only child in a household of adults. In addition to his parents this included his grandfather, grandmother, and great-aunt. He played alone, made up games, many of which seemed to revolve around mechanics. Once, after seeing the statue of a fawn on an aunt's parlor table, the boy for weeks made plans to create an entire estate of animals. "I was going to put machines in them so they would work," he later recalled. "The whole estate would be made up of automatons, though I did not know that word at the time. I thought of it by the hour. The grown-ups to whom I tried to tell the plan would not listen, except the hired man."

By the time he was seven, Willis would lose himself doing self-assigned problems in the evening, after he had finished his share of the farm work. One of the first problems he tackled was a perpetual-motion machine. When he was nine his arithmetic class in the one-room school at Evans Center reached fractions and Willis faced what seemed to him an insurmountable barrier. He could not grasp the meaning of fractions. When his mother noticed he was acting worried she found what was wrong and took action, which Carrier in later life called the most important thing that ever happened to him.

> My mother told me to go to the cellar and bring up a pan of apples. She had me cut them into halves, quarters, and eighths, and add and subtract the parts. Fractions took on meaning, and I was very proud. I felt as if I'd made a great discovery. No problems would be too hard for me after that—I'd simply break them down to something simple and then they would be easy to solve. From then on I worked on arithmetic far beyond my class assignments. Once I worked almost a year to get the answer to one problem, but it did not shake my faith in the method my mother had taught me. She opened up a new world to me and gave me a pattern for solving problems that I've followed ever since. In one-half hour she educated me!

Carrier always said that what talent he had in mechanics "I inherited from my mother." Once she fixed an alarm clock "and alarm clocks were not common then." At another time she told him about a paper mill she had visited as a girl. Her description was so accurate that years later, when Carrier made his first visit to a paper mill in order to design an air conditioning system, he felt as if he had been there before.

Willis's mother died when he was eleven years old. For a while his Aunt Abbey, who was his father's sister, lived with the family. Once in trying to help her with the churning, Willis upset a bucket of sour cream hanging in the well. When he and his father cleaned the well the pump was removed. Willis's aunt made an indelible impression upon him by explaining how the pump worked and stating that "the atmosphere exerts a pressure of about fifteen pounds per square inch." Years later Willis Carrier observed:

> That was the first time I ever heard of atmospheric pressure, but I did not realize then that air is elastic. That fifteen pounds pressure is a number I'd never forget even if I'd not made air my main interest through the years.

There are many other incidents in Willis's boyhood which evidenced a mechanical aptitude and a scientific approach to life. One schoolmate recalled that the boy, when helping his father prune grapevines, "worked geometry in the snow." There is also the story that when he was fourteen Willis fixed an old clock whose pendulum support had broken. Another anecdote relates that when a new thresher was delivered, knocked-down from the manufacturer, young Willis was equal to the job of assembling it.

When the farm work was finished in the fall of 1890, Willis entered Angola Academy, later called Angola High School. The next four or five years he called "the toughening period of my life." Each morning he was up at five o'clock. He and his father and a hired man milked twenty-four cows. Willis then delivered the cans of milk to the railway milk stop at Pike's Crossing by wagon or, if the snow was deep, by bobsled. He would then drive home, have breakfast, and walk a mile across the fields to school. A classmate recalled that "Willis used to play baseball, skate, swim in Lake Erie, and box with us boys almost every day; but when milking time came, he went home."

Willis Carrier was not a conformist even in those days. He

4

remembered his mother's advice: "Figure out things for yourself." Often, instead of doing assigned lessons he worked on problems from more advanced textbooks he found in the library of his home. When he ran out of problems he would make them up for himself. One such was a formula for determining the day of the week on which any date would fall in any century. He recalled:

> I had a way of being far behind the other children. When they brought in their home work with ten or more problems correctly solved, I'd only have one or two—but I'd know my answers were right and why.

Willis entered into his work, play, and study with enthusiasm and it never occurred to him that he could not succeed at anything if he tried hard enough. In recalling how much energy he spent on each game and each problem, he once observed: "my conviction overcame my natural laziness." When he wrote his high school graduation essay in 1894, he titled it "Circumstances the Mold; Man the Molder." Forty years later, as a guest speaker at Angola High School, he referred to this graduation essay:

> My thesis was that a man with power of will could make himself anything he wished no matter what the circumstances. I know better now. I am here tonight to recant. I never could be an expert golfer. I found this out by trial and error. That, too, is education—to learn where one lacks aptitude.

When Willis was graduated from Angola Academy the nation was in a financial depression. Milk had dropped to six cents a gallon and grapes to ten cents a basket. The family farm was mortgaged and Willis had to help out financially instead of going on to college. For nearly two years he taught in one-room schools, living at home and paying for his board and room. By 1896, his ambition to enter Cornell University seemed no closer to realiza-

5

tion than when he left the Academy. He had neither the money nor the necessary entrance requirements.

Willis Carrier overcame these obstacles by seizing upon an opportunity presented by his stepmother, Mrs. Eugenia Tifft Martin, whom his father had married when Willis was fifteen. A widow with three grown children, one of whom was a veterinarian in Buffalo, she arranged for Willis to attend high school in Buffalo by living with this son and earning his board and room by helping the family. In the fall of 1896 the farm boy moved to the city and entered what was then known as Central High School. The following spring, after competitive examinations, young Willis found himself the winner of a state scholarship which paid his tuition at Cornell University for four years. He still lacked funds for board, room, books, and clothing, but he borrowed enough to enter a tutoring school at Ithaca, staking everything on passing competitive examinations for one of the university scholarships which provided funds for worthy students. He took the examinations and got down to his last nickel before he learned that he was the winner of the H. B. Lord Scholarship which provided him with two hundred dollars for each of two years. Of the momentous windfall, Carrier later said:

> For the two years which the scholarship lasted, I was not too badly off. Of course I could not dress well. But I wore a clean collar every day—they were not the celluloid collars, either, that some of the boys wore. I did not feel sorry for myself because I had so little money, never felt inferior, never was conscious of being snubbed. I was happy to have my chance to study engineering.

Carrier got through Cornell by stretching his funds and taking on odd jobs of almost all types. He mowed lawns, tended furnaces, tried waiting on tables, and later became an agent for a boarding house. In his senior year he and another student formed a cooperative student laundry agency and made close to a thousand dollars each. Theirs was the first of the student "co-ops" that operate on many campuses today. Willis spent two summers selling stereoscopes and views; during another he worked on setting up plans for the laundry agency.

One college mate recalls seeing Carrier at Cornell "going across the campus, a dark-complexioned chap, notebook or papers under his arm, and always in a hurry, walking in long strides." He won medals in boxing and in cross-country running, rowed on the senior class crew, passed all subjects easily and with credit except for a "condition" in free-hand drawing when a sophomore.

Another friend of his college days recalls that "Carrier would start explaining an idea in class and he was soon so far ahead of us all in his thinking that not even the professor could keep up with him or understand him." And another contemporary observed that "Willis was essentially a man's man, but got on very well with the girls in his class—especially with one of them, Claire Seymour." Willis Carrier and Claire Seymour became engaged while students and were married slightly over a year after their graduation from Cornell University in June of 1901— Carrier receiving the degree of Mechanical Engineer in Electrical Engineering.

Carrier began specializing in electrical engineering during his final year and the subject of his senior thesis was the "Design and Construction of an Alternating Current Wave-Tracer." In 1950 he explained, "I had planned to study electricity when still in Angola and chose it because it was then a new art, much as electronics is today." As his graduation neared he thought of himself as ready to work in electrical engineering and wanted to get a position with the General Electric Company. However, a representative of the Buffalo Forge Company invited him and three other Cornell seniors to visit the company's home office for employment interviews.

Carrier hesitated about deserting his long-made plans to specialize in electricity. He weighed the pros and cons and, although he recognized that by living at home his salary would go further, he concluded that he could use his knowledge of electricity even though his prospective employers were engaged in the manufacture of blowers, exhausters, and heaters. He accepted the invitation to visit the home office of Buffalo Forge and around nine o'clock on a June morning in 1901 was on his way. Carrier later related:

> Although I had lived in Buffalo, I had never heard of the Buffalo Forge Company before and didn't know where the plant was located. So, when I took the Broadway streetcar for the plant, I asked a fine appearing young man who sat next to me if he knew where Mortimer Street was and said that I was going to the Buffalo Forge Company. He replied: "I will show you, as I am going there myself."
>
> This young man said his name was Irvine Lyle and that he was in the employ of the Buffalo Forge Company. He said that he had been selling for them out of Syracuse and was considering a transfer to New York City to take charge of the office there.

It was a strange fate which brought together the two men who afterward were to be instrumental in introducing air conditioning into the world.

II

WILLIS Carrier went to work for the Buffalo Forge Company on July 1, 1901, at a salary of ten dollars a week. His hours were from 8 in the morning to 6 in the evening except on Saturdays, when he worked from 8 A.M. to 1 P.M. His first work was at the drafting board and involved a heating plant for the Erie City Boiler Company. Subsequently, he worked on a system for drying lumber, another for drying coffee, and on a third which involved forced draft for boilers.

By the time Carrier had finished these first assignments, he realized that available data were insufficient to design soundly-engineered heating, drying, and forced draft systems. Because of the rule of thumb techniques which were common practice at the time, Carrier believed that engineers were allowing excessively large "factors of safety" in designing and installing equipment and considered that these were really "factors of ignorance."

No one suggested Carrier obtain more data and, in fact, few engineers recognized that those available were inadequate. But he posed the problem to himself and set about to find the answer. He did the necessary research after working hours. His first study involved the reading of much published material on mechanical draft. The result was a formula for selecting draft fans for maximum boiler efficiency with minimum fan horsepower.

9

Carrier described this study and the findings that resulted from it in a paper entitled "Mechanical Draft" which he presented at the Buffalo Forge Company's annual sales meeting in December of 1901. His paper was a highly theoretical one on a very practical subject and, although it was delivered by a young engineer who had been with the company less than six months, it impressed salesmen, experienced engineers, and the two top executives of the firm: William F. Wendt, the founder and principal owner, and his younger brother, Henry W. Wendt, the operating manager. As a result, the Wendts decided that Carrier's research should be continued, not after hours but during the regular working day. So the twenty-five-year-old engineer was granted permission to set up what later became an industrial laboratory.

The first research work undertaken had to do with heaters, since this field promised returns of immediate value to the company, engaged as it was in the manufacture of heating coils and the fans which circulated air over the coils. The only data then available were physical and thermal properties of steam and air. Unknown to engineers of that day was reliable information on how much heat air would absorb when it was circulated over steam-heating coils. To obtain the missing data, Carrier began experiments in the Buffalo Forge shop and shortly was assigned helpers to make slide rule calculations converting the data to curves and tables. One engineer, looking back on those days, recalled the energy with which Carrier pushed the tests out in the shop:

> Our desks in the main office were right next to each other. I was at that time greatly impressed by the activity of his mind. I remember quite well that he used to come in, sit down at his desk and work awhile, then get an idea and run out full speed for the shop, where he would make some changes or additions to his equipment. This would occur all day, every day.

When the tests were ended, the equations derived, and the calculations completed, Carrier had heater data of tremendous value to the company. The estimating engineer, for the first time, could read from a table for various steam pressures just how much each square foot of surface in a pipe coil heater would heat the air blown over it at various velocities and temperatures. But the data were not used immediately by the engineers. How to get them to use the tables was a problem that worried Carrier. Once, as he sat at his desk, gazing at the ceiling while pondering this problem, Carrier heard the scolding voice of Henry Wendt.

"Young man, you can't stay around here if you don't apply yourself," he said, and moved on into his own office.

Carrier was so angry he followed him into the office and said: "Mr. Wendt, what I was thinking about is worth more to this company than what can be done in three years of regular work."

Eventually the heater data were applied by the engineers and during the first heating season in which they were used they saved the company $40,000 which previously had been spent in correcting unsatisfactory installations. Eventually all of the industry felt the impact of Carrier's research and credited him with establishing the pattern for testing and rating heaters which is used today.

Even before Carrier's heater data had proved their worth, he suggested the company undertake a broad research program to provide adequate data in other fields so that better products could be built, more efficient systems designed, and customer complaints reduced. Henry Wendt carried the suggestion to his older brother, who asked Carrier to explain his idea. Years later Carrier told this story of the conversation:

> When Mr. W. F. asked me how much I thought the research program for one year would cost, I told him about three thousand dollars. He thought I had said thirty thousand dollars, and said it was too much. When he learned

it was just three thousand, it sounded so little in comparison that he gave consent for me to start research right away.

Thus, in the summer of 1902, a research program was inaugurated by the Buffalo Forge Company. Carrier directed the work without a title and, unless there was a lull in the engineering department, often without a staff. He had no laboratory and was compelled to set up his experimental apparatus in the shops wherever steam, water, and electricity were available. His salary was raised to twenty dollars a week and his research now had the prestige of a name: the Department of Experimental Engineering. It was the first research department set up in the heating and ventilating industry. Carrier's suggestion to establish it was accepted in July of 1902, a month which brought another event of much more fateful consequence.

III

WHEN Willis Carrier presented his paper at the Buffalo Forge sales meeting in December of 1901, one of the audience was the friendly person he had met on the streetcar the morning he applied for a position. This man was Irvine Lyle, a graduate engineer who had grown up on a farm in Woodford County, Kentucky. He had attended the University of Kentucky, played football there, and was graduated in 1896 with a degree in mechanical engineering. Lyle was six feet two inches tall, stood erect, usually spoke in a low voice, but could be very firm indeed. He was once described by Carrier as:

> The best salesman I ever knew. He was serious and hard working. People liked him and trusted him. Because he believed in what he sold, he made prospective customers believe in it too.

By the time of the sales meeting, Lyle was being transferred from Central New York with headquarters in Syracuse, to become manager of the New York office of the Buffalo Forge Company. Lyle heard Carrier read his paper. Apparently he was much impressed, for in the spring of 1902 he sent to the home office an inquiry from a New York consulting engineer with the request that the research work required be placed in Carrier's hands. The problem was essentially how to control humidity of air. It was a problem almost as old as man himself.

For five thousand years human beings had tried without success to conquer the discomfort caused by hot, humid air. The fan, whose origin is lost in antiquity, was probably the first of many such devices. One of the next recorded efforts involved evaporative cooling employed by an Assyrian merchant. Three thousand years before Christ he had the walls and floor of a room below his courtyard sprayed with water by his servants in hot weather. Several Roman emperors reportedly brought snow from the mountains to cool their gardens in the summer. Around 775 A.D. Caliph Mahdi of Baghdad had a summer residence built with double walls between which imported snow was packed.

The machine age intensified the heat problem. People who had once worked at home moved into factories where the heat of machines, fellow workers, and lights was added to the heat of the sun beating down on buildings. Thus men who had lived in a tolerable climate were frequently forced to labor in oppressive indoor weather.

Beginning in 1775, when Dr. William Cullen began research to create cold by evaporating a liquid, many men tried to manufacture ice. The first American to design an ice-making machine was Dr. John Gorrie, of Apalachicola, Florida, who was issued a patent in 1851. By the 1880's refrigeration was commercially accepted in the southern United States and its use began to spread throughout the country. A brewery was equipped with refrigerated coils over which air was drawn. A restaurant was cooled by embedding air pipes in ice and salt, then circulating the chilled air. Madison Square Theatre in New York City was equipped to use four tons of ice a night to keep its patrons comfortable. And a system was devised for Carnegie Hall which provided for ice held in racks at the outdoor air intakes, but was never used. Stirred by the question, "If they can cool dead hogs in Chicago, why not live bulls and bears on the New York Stock Exchange," the members decided to install air-cooling apparatus in their proposed new building. Alfred R. Wolff designed the system which operated successfully for twenty years.

During these many centuries, man had learned how to cool, circulate, and moisten air. Many of them had been aware of the phenomenon of dehumidification which occurred when air was cooled. But until the work of Carrier in 1902 no one had succeeded in reducing the humidity of air and holding the moisture content to a specified level. It remained for the young engineer, a year out of Cornell, to blaze a trail into the unknown.

The problem Lyle brought to Carrier in the spring of that year involved humidity in the plant of the Sackett-Wilhelms Lithographing and Publishing Company of Brooklyn, New York. Lithographers of that time, like many other manufacturers working with hygroscopic materials, were troubled by the weather. Atmospheric conditions caused paper to expand or contract. It was one size on hot, humid days, another size on hot, dry days. When printing in color, similar distressing changes often occurred between runs. Colors overlapped or failed to match those printed on another day. The flow of ink and its rate of drying were affected. The result was that printers often had to reprint jobs or drastically reduce the speed of their presses in order to maintain quality. Since Sackett-Wilhelms printed such publications as the humor magazine *Judge,* it was continuously confronted with deadlines. Consequently they could not afford to risk the hazards involved.

After the extremely hot summers of 1900 and 1901, the publishing company called in Walter W. Timmis, a New York consulting engineer. Timmis took the Sackett-Wilhelms production problem to Lyle, who turned it over to Carrier and provided him with U. S. Weather Bureau tables, the only reliable data on psychrometrics in existence at the time. These tables were based on empirical formulae, later questioned by Carrier when he devised his rational psychrometric formulae. With the Weather Bureau tables, and plans and specifications of the lithographing plant, the twenty-five-year-old engineer started tests on dehumidifying air.

Willis Carrier apparently never questioned the possibility of a solution to the problem nor his ability to find it. He had worked with fans, heaters, and temperature control, but had had no occasion to consider the control of the relative humidity of air. He acted characteristically—quickly grasped the basic factors involved, reduced them to their simplest terms, and planned two tests. One was based on the suggestion from Timmis, the other on his own idea. Later Carrier recalled:

> To try out the suggestion from Mr. Timmis, we rigged up a roller towel arrangement with loosely woven burlap which we kept flooded with a saturated solution of calcium chloride brine. We drew air through the burlap with a fan. Readings of dry-bulb and wet-bulb temperatures on both sides of the brine-soaked burlap told us the amount of moisture removed from the air by the brine.
>
> Everything about the test was operated manually except the fan. It would be a short test, I thought, not worth requisitioning a motor to roll the burlap and a pump and motor to circulate the brine. Manpower was cheaper for a short test—a man to dip brine from a barrel and pour it over the cloth, and a man to turn the rollers—especially when manpower cost ten or twelve dollars per week per man.

The job of dipping the brine was assigned to Edward T. Murphy, who later was to become a founder and senior vice-president of Carrier Corporation. Another engineer turned the rollers and Carrier himself took the temperature readings. In about a week the test was abandoned. The calcium chloride took the moisture out of the air but the air was salty as it left the apparatus—a condition that would cause machinery to rust.

Carrier then turned to his own idea for dehumidifying air and holding its moisture content constant. For his tests he converted apparatus with which he had tested heaters and, in place of steam, circulated cold water through the heating coils. He selected from the Weather Bureau tables the dew-point temperature at which the air has the right amount of moisture for the printing process. He then set about balancing the temperature of the coil surface and the rate of air flow to pull the air temperature down to the selected dew-point temperature. When he completed the tests, he knew the amount of coils, their temperatures, and the necessary air velocities to give the printers the kind and volume of air specified. From the data he specified the equipment for the Sackett-Wilhelms installation. On July 17, 1902 drawings were completed for what came to be recognized as the world's first scientific air conditioning system.

The 1902 installation marked the birth of the air conditioning industry because of the addition of humidity control. Authorities in the field now recognize that air conditioning must perform four basic functions: (1) control of temperature, (2) control of humidity, (3) control of air circulation and ventilation, and (4) cleansing of the air.

The classic definition as given by Dr. Carrier some years later is:

> Air conditioning is the control of the humidity of air by either increasing or decreasing its moisture content. Added to the control of humidity are the control of temperature by either heating or cooling the air, the purification of the air by washing or filtering the air, and the control of air motion and ventilation.

Carrier learned from his tests what the air conditioning engineer of today selects from tables—the amount and temperature of chilled water and the rate of flow through the coil required to cool and dehumidify each cubic foot of air to a specified temperature and relative humidity.

Sackett-Wilhelms had specified an indoor temperature of 70 degrees F in winter and 80 degrees F in summer, and a relative humidity of 55 percent the year around. For the winter condition, Carrier and Lyle decided to supplement the air-heating coils with a humidifier consisting of perforated pipes supplied by low-pressure steam from the plant boilers. To maintain 55 percent humidity on a zero day in winter they calculated they would have to add 86 gallons of moisture every hour to the 20,000 cubic feet of ventilation air brought into the building—a steam requirement of 24 boiler horsepower.

The summer situation was just the reverse. It was necessary to furnish cool air to absorb the large quantities of heat from lights, printing presses, occupants, and leakage from outdoors. In addition, there were large quantities of moisture—50 pounds per hour from the pressmen, 45 pounds from the drying of the ink, and about 220 pounds from the humid ventilation air—a total of 400 pounds of water per hour which the air conditioning system had to remove in order to maintain the desired indoor conditions. The cooling and dehumidification were accomplished by two sections of cooling coil. One used cold water from an artesian well, the other was connected to an ammonia refrigerating machine. Taken together, their cooling effect totaled 54 tons, the equivalent of melting 108,000 pounds of ice in a 24-hour day. The installation was indeed a milestone in man's control of his indoor climate!

Carrier and Lyle recognized that low operating costs on this first installation would certainly influence future sales. Consequently they did not draw from the city water supply then available but re-used the two hundred gallons of water per minute from an artesian well. After this water left the first set of coils, it served as jacket water and condensing water for the refrigerating system and was then pumped to a tank on the roof to serve the plant's plumbing system.

The first set of coils was installed late in the summer of 1902 along with fans, ducts, heaters, perforated steam pipes for humidification, and temperature controls. The refrigerating system, using a De La Vergne ammonia compressor, was added in the spring of 1903 to meet the first full summer's operation of the world's initial scientific air conditioning installation. On October 21, 1903, in a letter to his home office, Lyle reported that "the cooling coils which we sold this company have given excellent results during the past summer." Thus, out of Willis Carrier's research and ingenuity and Irvine Lyle's faith and salesmanship, a new industry was conceived and given birth.

IV

WILLIS Carrier was not content with his first air conditioning system. He was confident that a more practical method for dehumidifying air could be found. And his mind, as he later said, "kept returning to the studies on air. I was not satisfied with the answer I'd found for the lithographing plant." He wrestled with the problem of controlling the moisture in air while en route to and from the plant, at mealtimes, while shaving, even when people were talking to him, and especially when traveling.

On one trip in the late fall of 1902, Carrier had to wait for a train in Pittsburgh. It was evening, the temperature was in the low thirties, and the railway platform was wrapped in a dense fog. As Carrier paced back and forth, waiting for his train, he began thinking about fog. As he thought he got the "flash of genius," as patent experts put it, that eventually resulted in "dew-point control," which became the fundamental basis of the entire air conditioning industry. Carrier's conception that historic evening in Pittsburgh, as dictated by him years later and read and corrected by him a few months before his death, went along these lines:

> Here is air approximately 100 percent saturated with moisture. The temperature is low so, even though saturated, there is not much actual moisture. There could not be at so low a temperature. Now, if I can saturate air and control its temperature at saturation, I can get air with any

amount of moisture I want in it. I can do it, too, by drawing the air through a fine spray of water to create actual fog. By controlling the water temperature I can control the temperature at saturation. When very moist air is desired, I'll heat the water. When very dry air is desired, that is, air with a small amount of moisture, I'll use cold water to get low temperature saturation. The cold spray water will actually be the condensing surface. I certainly will get rid of the rusting difficulties that occur when using steel coils for condensing vapor in air. Water won't rust.

When Carrier returned to Buffalo he did not immediately start research on equipment to create a fog to saturate air. Other more pressing studies were under way and even they were frequently delayed by inspection trips. It was not until the spring of 1903 that he had an opportunity to concentrate on applying his fog theory to practice. He investigated air washers then on the market, found them satisfactory for cleaning air but inefficient for saturating it. Later he described his mechanical problem as follows:

I found I'd have to design a new apparatus. It must produce a very fine spray, actually a mist, in order to completely saturate the air. Even in some rainstorms the air may not be saturated—may have a relative humidity of 90 percent or less instead of 100 percent for saturation. My apparatus had to do a better job than rain; that meant a mistlike spray and I had a hard time getting a nozzle to produce the fine spray.

Also, I had to make sure that no unevaporated particles of water were carried in the air stream after it passed through the mist of water. The particles would evaporate later, raise the moisture content of the air. To prevent entrained moisture I needed eliminators so efficient that no moisture would get through.

21

Carrier's work on apparatus to produce a fog mechanically was intermittent. Meanwhile, in his studies of fan-heater drying systems he made observations and calculations concerning air which were fundamental to air conditioning. From his first tests on air, he believed that the wet-bulb temperature of the air remained constant when water evaporated into it. As he once expressed to a non-technical listener:

> Air, as we find it in nature, is always a mixture of dry air and water vapor. The dry air contains heat, sensible heat; the vapor contains heat, latent heat. Total heat of the mixture is the sum of the two. I believed from my first tests on air that, when the moisture in air is changed, there is a transfer of sensible heat to latent heat or the reverse, but always the total heat of the mixture remained unchanged.

His chance to test this hypothesis came before he designed his apparatus to saturate air. The Eastern Tanners Glue Company, at Gowanda, New York, had installed a Buffalo Forge fan and heater for its glue-drying tunnel. When the plant superintendent asked for an engineer to check the fan performance, Carrier was sent. He took along the usual fan-testing instruments plus two wet-bulb and two dry-bulb thermometers. After he had tested the fan and found that it delivered its rated capacity of air, he started some tests of his own, which he later described as follows:

> The wet glue entered one end of the tunnel and left at the other. A fan blew air into the dry or leaving end. At each end I hung a wet- and dry-bulb thermometer. As I recall, the air entered at about 90 degrees F dry-bulb and 60 degrees F wet-bulb. It left at about 62 degrees F dry-bulb and 60 degrees F wet-bulb. The wet-bulb remained constant throughout the entire length of the tunnel—the air absorbing moisture all the time. The dry-bulb temperature dropped, the sensible heat grew less, and the latent

heat increased by the same amount—the total heat of the air and the vapor remained constant. My reasoning had been correct. I had proved my theory to back up my fog idea for controlling moisture in air.

By late 1903 Carrier had designed an efficient eliminator to prevent moisture particles from entering the air stream as it passed from the fog chamber. But he had not succeeded in creating mechanically the proper fog. He was still looking for a nozzle that would produce a fine mist and handle a large quantity of water at the same time when Professor W. N. Barnard, of the Sibley College of Engineering at Cornell University, called on the Buffalo Forge Company to see how his former students were making out. When Carrier spoke of his search, Barnard told him of a nozzle patented by his father, a horticulturist, which he had devised for spraying insecticides. Carrier secured several of these nozzles, modified them to handle water, and made royalty arrangements. Now, with the proper nozzle and an efficient eliminator, he quickly completed the mechanism he had visualized two years before on the Pittsburgh railroad platform. On September 16, 1904, Carrier applied for a patent on his invention, which he called an "Apparatus for Treating Air." The patent, No. 808897, was issued on January 2, 1906.

Carrier's "Apparatus for Treating Air" was the world's first spray-type air conditioning equipment. It was designed to humidify or dehumidify air, heating water for the first and cooling it for the second. The use of spray water for humidifying was readily accepted, but Carrier's idea of dehumidifying air by using water was so revolutionary that it was greeted with incredulity and, in some cases, with ridicule. However, Carrier proved air could be dried with water and two years after having designed the first scientific air conditioning installation, he provided a new industry with a new device which was to open thousands of industrial doors.

V

THE first sale of Carrier's "Apparatus for Treating Air" was made late in 1904 to the LaCrosse National Bank, of LaCrosse, Wisconsin. But the apparatus was used only to wash air in the ventilating system. And this illustrates why many people in the heating and ventilating industry did not thoroughly understand the revolutionary nature of Carrier's discovery and invention.

While Carrier welcomed any sale that would increase volume, he looked to a wider market: industries which, for efficient production, needed to control the moisture in the air. To help sales engineers reach this market, Carrier began writing a catalogue in 1905 which was published the following year under the title "Buffalo Air Washer and Humidifier." In the catalogue Carrier published data not found in textbooks of the time, defined psychrometric terms, and included a hygrometric chart which, when refined and published in 1911, was to bring him international fame.

In the pages of his catalogue, Carrier described the nozzle which caused the water "to burst into an almost invisible mist" across the entire chamber through which the air passed. He explained how the eliminator was made up of plates arranged to form "a series of unbroken sinuous passageways for the air." More significantly, he listed markets for the apparatus: cooling theaters, churches, restaurants, ships' stores; the industrial drying of soap, leather, glue; humidity control in textile mills and tobacco warehouses.

"The average cooling effect produced by the air washer in summer and without mechanical refrigeration is about 10 degrees," Carrier wrote. He explained that the effect resulted from the evaporation of the spray water and added that, for low temperatures and low moisture-content air "a refrigerating machine is used to cool the spray water to the desired point."

While Carrier was writing the catalogue he saw that the industry needed a handbook from which engineers could get pertinent data on air, and how to control it, without referring to numerous books and obscure articles. It took Carrier eight years to compile the handbook. Buffalo Forge published the first edition in 1914 and the fifth edition in 1948.

In 1905 Carrier was made head of the Buffalo Forge Engineering Department. And he was only twenty-nine. From then on he filled two positions with the company, directing research and supervising all engineering. He combined the two positions advantageously. When there was a lull in the Engineering Department the men worked on research, running tests or tabulating data from earlier experiments. The research department got its own building, a one-story structure under the water tower across the street from the shops. Here Carrier set up permanent test equipment, including apparatus to study air flow in ducts.

At that time air ducts were built large and round—large to make sure they would permit easy flow of the desired volume, round because that shape simplified fabrication. Many dampers were installed to block off runs which received more air than specified. "It was like tailoring all suits for unusually large men and then cutting each down to customer size," Carrier explained. "Very wasteful."

Carrier believed that distributing systems could be designed so that each run would receive its allotted amount of air automatically—like water seeking its own level. In this way he thought he could dispense with damper blocking. He tested straight ducts, elbows, tees, and take-offs. He devised a formula to calculate the sizes of all branches so that each offered the same resistance to flow. He called his method "Proportioning Piping for Constant Friction." Later he developed a formula for designing rectangular ducts with constant friction. Carrier's research on ducts provided data for improved design of air-distributing systems and established a pattern adopted by the entire industry for fan-heating air-distributing systems—a pattern which nearly half a century later is still being recommended as one of three means for efficient duct design.

Meanwhile, the "Apparatus for Treating Air" was given more tests by two engineers working under Carrier's direction. One of these was I. H. Hardeman, a graduate of Georgia Institute of Technology where he had specialized in textile engineering. Hardeman recognized the apparatus as applicable to textile mills, which required moist air for efficient production. He supposedly told Carrier, "Revolutionize the textile industry—that's what your equipment is going to do." In any event, Carrier learned much about textile mills from Hardeman, caught some of his enthusiasm, and in April of 1906, without having ever been in a textile mill, published an article in *Textile World* in which he stressed the advantages of his apparatus for "maintaining the proper temperatures and humidity."

By the time this article appeared, two events of historic importance had taken place. First, Stuart W. Cramer, of Charlotte, North Carolina, a leading textile engineer, had given the infant air conditioning industry its name. He used the term "air conditioning" in a patent claim filed for a humidifying head in April of 1906. The next month he defined the expression in an address before a convention of cotton manufacturers: "I have used the term 'air conditioning' to include humidifying and air cleansing and heating and ventilation."

The second historic event of early 1906 was the sale by Hardeman, now the Buffalo Forge representative in the South, of Carrier's "Apparatus for Treating Air" to the Chronicle Cotton Mills in Belmont, North Carolina. The apparatus was added to the mill's fan-heater ventilating system. After the installation all the air circulated in the mill, whether drawn from outdoors or from the mill as returned air, passed through the water spray and eliminator into the fan. It was the first industrial "central station" humidifying system.

Carrier himself visited the Belmont installation in the summer of 1906, on his way back to Buffalo from New Orleans where he had gone to correct a pump failure in a sugar refinery. His visit had a great impact on both the textile and the air conditioning industries. Later he described this first visit to a textile mill:

> When I saw 5000 spindles spinning so fast and getting so hot that they'd cause a bad burn when touched several minutes after shutdown, I realized our humidifier was too small for the job. All of the heat played havoc with relative humidity—raised air temperatures far beyond what we had calculated.

Some textile men of the time believed that the heat of the machines went into twisting the threads. Carrier realized that practically all of the power employed in driving the machinery ultimately was converted into heat within the building. He found "the heat of the machines alone was sufficient to raise the dry-bulb temperature 33 degrees F which meant, if constant humidity was to be maintained, that more moisture must be added to the air." He later added:

> I had no choice but to run up the water pressure. Inasmuch as the fan fixed the air volume, I could not circulate a great quantity of humid air so I circulated the same quantity of air, but each cubic foot contained much more moisture than originally planned. I had to carry a high dew-

point, higher than the humidifier could supply under normal operation. I kept running up the water pressure on the nozzles, from 25 pounds to 50 pounds to 100 pounds. At 100 pounds I got enough moisture in the air to hold the card room around 62 percent relative humidity, the spinning room around 70 percent, and the winding room around 85 percent.

Carrier concluded that he must change his spray apparatus—somehow add more moisture to the air without pushing pump pressures to 100 pounds a square inch or appreciably increasing the size of the pump. By this time several fan-heating-humidifying systems had been installed elsewhere. They were designed to humidify, heat, and ventilate, whereas the Chronicle Mills installation involved humidifying adapted to an existing heating and ventilating system. As a result the engineers under Carrier saw that the air volumes in the new systems were adequate for all three objectives. The performance of the systems satisfied customers, but Carrier was determined to improve the spray apparatus. He did it by turning the nozzles around. Instead of discharging the water downstream with the air, he discharged it against the air stream. The counter-flow impact produced higher moisture absorption rates with lower water pressures on the nozzles.

Carrier now had two types of apparatus: one with downstream sprays, which was sold as an air-washer for installations not requiring moisture control; the other with counter-flow sprays, which was sold wherever the moisture in the air was to be controlled for humidifying or dehumidifying. The first counter-flow spray apparatus was installed early in 1907 at the Lowell Cotton Mills, at Lowell, North Carolina. When checked, its performance was found satisfactory with water pressure at the nozzle of 25 pounds per square inch.

Carrier's visit to the Chronicle Mills in 1906 also led to a discovery fundamental to the air conditioning industry. As he took readings in the hot mill he found that, within a wide limit, the relative humidity of air remained constant as long as the difference between the dry-bulb temperature and the dew-point temperature remained constant. If he saturated air at 16 degrees F below room temperature, he was able to hold 60 percent relative humidity whether the thermometer read 80 degrees F (dew point 65 F) or 90 degrees F (dew point 74 F). Carrier's discovery, that "constant dew-point depression provided practically constant relative humidity," later became known among air conditioning engineers as the "law of constant dew-point depression." On this discovery he based the design of an automatic control system for which he filed a patent claim on May 17, 1907. The patent, No. 1,085,971, was issued on February 3, 1914. Carrier was thereby recognized as the inventor of what became known as "dew-point control."

While Carrier was developing his first automatic control system, the infant air conditioning industry was making an epochal expansion; from the North Carolina cotton mill to a Michigan pharmaceutical plant. Buffalo Forge's sales manager, Charles Arthur Booth, had interested Parke, Davis & Company of Detroit in an air-cooling and dehumidifying system for the room in which capsules were made. Booth and Carrier visited the plant, surveyed the room, and learned that capsule making was vulnerable to both humidity and heat. In warm air, the gelatine flowed unevenly, so that the thickness of capsule walls varied; in humid air the drying slowed up, so that production suffered. They learned that the capsule-forming machines operated satisfactorily if the temperature did not exceed 80 degrees F and the relative humidity 60 percent.

Back in Buffalo, Carrier laid out an air-cooling and dehumidifying system. It was guaranteed to maintain the specified 80 degrees F with 60 percent relative humidity when outdoor temperatures were as high as 90 degrees. At the same time the system took care of the indoor evaporation of 137 pounds of moisture from the gelatine forming the capsules. The contract price of $16,917 included a fan, heaters, ducts, dehumidifier, and a 25-ton ammonia refrigerating machine to chill the spray water.

When Parke, Davis signed the contract, both Carrier and Booth were elated—Carrier at an opportunity to try out his low temperature fog idea, Booth at the prospect of the sale opening new markets for the company's products. They presented the contract to W. F. Wendt for approval.

"How many people work in the Capsule Department?" he asked.

"Eighty-five."

"If there's a leak in the ammonia refrigerating coils," he continued, "will the spray water absorb the gas and release it in the air blown into the room? Wouldn't that endanger the workers?"

When the last two questions were answered affirmatively, Wendt said, "I'll not accept the contract."

He explained that the Buffalo Forge Company did not make the refrigerating equipment, hence could not be sure of its workmanship and safety, nor afford to take a chance. Years later, Carrier said:

> But Parke, Davis wanted the air conditioning system, asked for and received permission to use our drawings and specifications, bought our fan, heaters, and dehumidifier, purchased a refrigerating machine from Great Lakes Refrigerating Company, and installed the system. It worked all right, too. No ammonia leaks caused trouble.

Thus, Parke, Davis & Company installed the first low temperature spray-type air conditioning system. Before the end of 1907 Carrier's air-treating apparatus had also been sold to additional cotton mills, a worsted mill, a shoe factory, and two silk mills, including the Fugi Silk Spinning Company in Yokohama, Japan.

A sale to a silk mill in Wayland, New York was particularly

significant. This plant had been built in 1906 to weave silk chiffon, a fabric so soft and fine that its full width can be pulled through a finger ring. In the first year of operation it was found that without control of humidity, the cloth could not be produced free from shadows caused by uneven threads and uneven weaving. A sales representative working under Irvine Lyle in Buffalo Forge's New York office heard of the difficulties, surveyed the weave room, and dispatched his figures and sketch to Carrier.

When Carrier began designing a system for the Wayland mill, he was determined to have no underestimation of the amount of heat involved, as had occurred at the Chronicle Mills in North Carolina. He considered every possible source of heat affecting the mill air, including one up to then ignored by engineers: the heat of the sun on the building itself. The task of finding how much heat was added by the sun was turned over by Carrier to his young assistant, A. E. Stacey, Jr., who had come to Buffalo Forge the year before from Syracuse University. Stacey took his task to the library, where about all he could find was the heat gain from the sun in the Sahara. But Carrier made assumptions, arrived at a figure to use in calculating heat gain from the sun at Wayland. The proposal dated May 27, 1907, included the following clauses:

> We guarantee the apparatus we propose to furnish you to be capable of heating your mill to a temperature of 70 degrees F when outside temperature is not lower than 10 degrees F below zero.
>
> We also guarantee you that by means of an adjustable automatic control it will enable you to vary the humidity with varying temperatures and enable you to get any humidity up to 85 percent with 70 degrees F in the mill in the winter.
>
> In summertime we guarantee that you will be able to obtain 75 percent humidity in the mill without increasing

the temperature above outside temperature. Or, that you may be able to get 85 percent in the mill with an increase in temperature of approximately 5 degrees F above outside temperature.

The contract with the Wayland mill amounting to $3,100 was signed May 28, 1907. Stacey made the drawings. The heaters, humidifiers, fan, and duct work were installed under the direction of Edmund P. Heckel, another young engineer who had come to Buffalo Forge in 1905. Stacey then installed the controls and, in December of 1907, went to Wayland with Carrier to inspect the operation. Thus, four of the seven men who later were to grasp opportunity and form the Carrier Engineering Corporation—Carrier, Lyle, Stacey, and Heckel—were all involved in this one system. It was designed with the added heat of the sun taken into consideration and it was the first automatically-operated modern air conditioning system.

VI

AFTER his 1907 vacation, spent camping in the Ontario wilds where he had time to think as well as fish, Carrier got the idea that Buffalo Forge should set up a subsidiary company to engineer and market air conditioning systems. Charles Arthur Booth viewed the idea conservatively, but Irvine Lyle was enthusiastic about it. When the proposal was carried to the Wendts they quickly approved and, around November 1, 1907, drew up the first papers for a subsidiary company. At Lyle's suggestion it was named Carrier Air Conditioning Company of America, was wholly owned by the Buffalo Forge Company, and was in actual operation by early 1908. William F. Wendt was president, Henry W. Wendt, treasurer, and Willis H. Carrier, vice-president. Carrier years later explained:

> The reason, I believe, "Mr. W. F." did not use the parent company's name was that, in case the company was not successful, there would be no stigma to the name of the Buffalo Forge Company. There were various reasons for the name of the new company. Among others, our competitors would be less hesitant to sell us fans than if the tie-in with Buffalo Forge was emphasized. And as to "Air Conditioning" in our title, it did not take us long to adopt Mr. Cramer's expression for controlled humidity of air. The phrase "of America" probably expressed Lyle's

33

hope in the company. If all went well, there'd be other subsidiaries—of Europe, of Asia.

Although the new company was named for Carrier, he continued as Buffalo Forge's chief engineer and director of research. The parent company agreed to furnish the subsidiary with free engineering consultations, but Carrier Air Conditioning was to pay Willis Carrier's expenses and also pay Buffalo Forge for his time at the rate of fifteen dollars a day whenever he was out of Buffalo on business for Carrier Air Conditioning. The sales manager of the new company, with headquarters in New York, was Irvine Lyle, and the construction superintendent was Edmund P. Heckel. Lyle's staff included: in Boston, his brother, Ernest T. Lyle, an experienced salesman for Buffalo Forge; and, in Philadelphia, E. T. Murphy, who back in 1902 had worked with Carrier and Lyle on the first air conditioning installation. In New York, A. E. Stacey served as chief engineer. When he was transferred to Chicago in 1909, L. Logan Lewis joined the company as chief engineer, later becoming a founder and vice president of Carrier Corporation.

The ink was hardly dry on the new company's letterheads when Irvine Lyle brought in an $8,000 contract. Lyle's customer was the Celluloid Company, which later became the Plastics Division of Celanese Corporation of America. The company had run into difficulties in its Newark plant, where celluloid film was made for the then infant motion-picture industry. During humid weather white specks formed which appeared as white spots on the screen when the film was magnified by the projector. Motion-picture producers naturally objected and the film manufacturer realized that something must be done. Lyle persuaded him to air condition the film base processing room. A memo by Carrier says:

> We went to work immediately on Lyle's contract to design the dehumidifying system. We worked under pressure,

anxious to make the most of a wonderful opportunity—a chance to introduce air conditioning in the film industry. The prospect for more film air conditioning contracts looked good, but we knew more film business depended on this first job being a success. Every piece of equipment that we considered using, we tested. Often we'd remain all night testing, checking, re-checking, and rigging up new equipment.

Manufacturers of the day viewed air conditioning with skepticism and considered the prices exorbitant. To keep costs down for Celluloid, Carrier designed a piece of equipment called a regenerator or heat exchanger which was used in air conditioning for a decade before becoming obsolete. The regenerator carried 10 to 15 percent of the air-cooling load. The remaining load involved reduction of the moisture in the air from $9\frac{1}{2}$ grains a cubic foot to $3\frac{1}{2}$ grains—i.e., to a dew point of 46 degrees F. In order to economize on refrigeration and pumping power, he designed a three-stage dehumidifier. In the first stage he circulated well-water which produced partial dehumidification, in the second and third stages he used water refrigerated in a room filled with ammonia coils. Here, installed in Newark, New Jersey in 1908, was the first spray-type apparatus built with more than one bank of nozzles in series. It enabled the Celluloid Company to produce film base without white specks. Then something went wrong and Carrier hurried to investigate. He traced the trouble to algae and quickly removed it by shooting steam through the heat exchanger. The result was three repeat orders for air conditioning from the same company.

Parenthetically, the Celluloid plant was the scene of one of many classic examples of Carrier's absent-mindedness. After working on the heat exchanger for a while one day, he remarked that it must be time for breakfast. Much to his surprise he learned that it was lunch time and that he had missed breakfast completely. On another occasion, during an inspection trip to Pennsylvania, he arrived with a large suitcase in which he had packed one handkerchief. At another time Carrier and Murphy visited a New England mill, looked over the system, then went to a restaurant for lunch. All through the meal Carrier drew diagrams on the tablecloth. The waitress would serve him one course, wait and wait for him to eat it, then remove the untouched food and serve the next course. Carrier left the dining room without realizing that he had eaten no food.

In those early days all the men in the Carrier Air Conditioning Company did all kinds of work—designing, testing, drawing plans, estimating jobs, selling contracts, testing again. One of the busiest of them all was Irvine Lyle. After the Celluloid contract, he divided his day into two types of activities: visiting consulting engineers, architects, and others to sell them the idea of air conditioning; and selling air conditioning systems themselves. Lyle also wrote catalogues for the new company. He turned out copy on trains, at night in hotel rooms, and over weekends at home.

Among the contracts Lyle signed in 1908 was one to air condition three floors of an American Thread Company mill at Fall River, Massachusetts, and a Farr Alpaca Company plant at Holyoke, Massachusetts. He also sold the Duplan Silk Company a system for its Hazelton, Pennsylvania, million-cubic-foot weave shed. This system used 1,000 gallons per minute of 52 degree F well-water. It was the first to employ well-water alone for low dew-point air and also the first to use an electric motor for the fan drive.

In 1908, Carrier was working on and completing six inventions to the patent application stage, two of them relating to automatic controls. He wanted the operation of an air conditioning system "to be independent of anyone's memory." At the same time, Lyle wanted spray apparatus standardized instead of being built to order and, to that end, had the factory tabulate for catalogue purposes the dimensions of standard air washers and humidifiers up to about 130,000 cubic feet per minute.

When the first year of Carrier Air Conditioning Company of America drew to an end the roots of the air conditioning industry which Carrier, Lyle, and their associates had planted had taken hold. These men, armed with more apparatus, satisfied customers, and sales literature, had equipped themselves to carry air conditioning into still more industries.

During the next five years, Carrier Air Conditioning Company invaded industry after industry. Irvine Lyle, remembering his boyhood on a Kentucky farm, visualized the importance of controlling moisture in tobacco and foresaw that industry as a great market for air conditioning. Early in 1909 he approached Thomas Hodge, a tobacco exporter in Henderson, Kentucky, showed him how air conditioning could enable accurate weighing and pricing, and sold him a $1,850 system to control the amount of moisture in his tobacco.

Two years later Lyle learned that a cigar machine, on which the American Tobacco Company had spent thousands of dollars, was not working properly. He figured the difficulty was probably the result of the tobacco being either too dry or too moist for the machine to handle. With Carrier, Lyle visited the Newark cigar plant which had bought the first machine from American Tobacco's subsidiary, the International Cigar Machine Company. Carrier designed a system to hold the cigar machine room at 80 percent relative humidity and the filler store room at 90 percent. He specified an interesting new measure to prevent fluctuating conditions: white blinds on windows to serve as reflectors and level out the effects of sun heat. The air conditioning system made mass production possible in the cigar plant, proved useful in further perfecting the machine, and led to an arrangement whereby the International Cigar Machine Company secured rights to sell Carrier equipment to tobacco plants throughout the world. The first sale, for $6,280, was made in 1913 following a visit by Carrier to an American Tobacco Company stemming room in Richmond, Virginia. He later said:

I never saw such a dusty atmosphere. I could see only a few feet in front of me, could not tell whether a person a few feet away was white or black, and could not see the windows across the room even when sunshine fell on them. All the workers wore handkerchiefs tied over their mouths and were instructed to breath through their mouths when in the room. Tobacco was stemmed by hand in those days and the process kept adding tobacco dust to the air. Later the process was changed from hand to thrashing in enclosed machines and that did away with the dust. But with hand stemming the dust was a problem, a big problem.

How to lay the dust? If we blew the air into the room as we did in most air conditioning installations, we would just stir up more dust. I figured the air introduced in the room must have a blanket effect. Then came the question of how to introduce a large quantity of air in a room to get such an effect. Back in Buffalo I fixed up a test to get the answer and from the test came the pan outlet.

This set-up which resulted in the pan outlet consisted of an air discharge opening placed in the ceiling of the test room. A flat rectangular pan, larger than the opening, was placed horizontally several inches below the ceiling. The air discharged from the opening hit the pan, spilled over its sides, and settled slowly to the floor. To prevent an induced air flow from the space under the center of the pan, I cut a small hole to let some of the air pass directly downward through the pan. It worked fine. We got the blanket effect all right.

Then we put the humidifying system in the stemming room of the tobacco plant in Richmond and distributed the air with pan outlets. The results were wonderful. When we started up the apparatus you could see the cloud of dust move to the floor, just as if it were a liquid being drained off from the bottom.

We got the blanket effect even though we introduced a sufficient volume of air to provide a three minute air change. That's a lot of air. It was high humidity air which caused the dust to absorb moisture, gain weight, and settle to the floor.

Everyone at the tobacco plant was amazed at the clean air in the stemming room and even Carrier was a little surprised at the speed with which the air conditioning settled the dust. But probably the most surprised was an old Negro janitor who, on learning what Carrier hoped to do, had said: "No use a dem foolin' their time 'way like dat—no one but da good Lord could stop dat dust." Carrier's system not only stopped it but the spray system also cooled the air by evaporation. The stemming room became a desirable place in which to work, and employees in other parts of the plant began coming in at lunch time to eat in the cooled clean air. A further result of Carrier's test connected with the pan outlet came, many years later, in the downward draft discharge of air so indispensable to comfortable air conditioning of theaters.

In 1908 Lyle sold the Astor Hotel in New York City a humidifier to remove dust and dirt from the air in one of its dining rooms. When on one warm day the apparatus lowered the indoor temperature ten degrees by evaporative cooling, the hotel ordered a second system for its Indian Room. These systems did not include mechanical refrigeration. In 1911, however, Kroeschell Brothers Ice Machine Company (subsequently Brunswick-Kroeschell Company which in 1930 was to become part of Carrier Corporation) bought an air washer, combined it with Kroeschell refrigerated coils, and obtained year-round air cooling in the

Pompeiian Dining Room of the Congress Hotel in Chicago. In 1914 both the Baltimore and Muehlebach hotels in Kansas City, Missouri, bought and installed Carrier air conditioning systems which included mechanical refrigeration.

But the big push of Carrier Air Conditioning Company was into industries where the immediate concern was products and processes, not personnel. In 1909 Irvine Lyle and Edward T. Murphy began investigating the steel industry, which seemed a potentially tremendous market for air conditioning because of the excessive fuel required to heat the moisture in the air blown through blast furnaces. Steel men were already saving fuel by removing this moisture, but Lyle and Murphy concluded Carrier's equipment could do the job more efficiently and at much less cost. Carrier began research on the problem, which involved determining psychrometric laws of high-pressure air. In two years a system was installed for the Northern Iron Company at Standish, New York, but savings fell below expectations.

It was a quarter of a century later—after the crude blowing engines of those pioneer days had been refined and improved—that another Carrier air conditioning system was sold to the steel industry to dehumidify blast-furnace air. However, Carrier's work on high-pressure air in 1909 enabled him, nine years later, to design a system to dehumidify compressed air for an automatic machine for the Gillette Safety Razor Company. This stopped the formation of rust spots on razor blades caused by moisture deposits during the wrapping.

By 1911 the Carrier Air Conditioning Company had moved into the rayon industry, which without air conditioning could have never achieved year-round, large-scale production. The company had also sold air conditioning to nine bakeries, a Michigan flour mill, a Youngstown rubber factory, and a Pittsburgh hospital ward for premature babies. Carrier, irked at learning that many purchasers of the air conditioning systems were using a competitor's fan, had taken time out to design, build, and test a new fan which "gave a wide range of capacity at a constant static pressure with but little variation in speed and with very little change in total efficiency." Called the "Niagara Conoidal," this multi-bladed forward-curved centrifugal fan was first shown to Buffalo Forge Company engineers in December of 1911 and was built in large quantities for many years thereafter.

In that same month Willis Carrier presented the most significant and epochal document ever prepared on air conditioning—his "Rational Psychrometric Formulae." Carrier read this paper, which has been called the Magna Carta of Psychrometrics, on December 3, 1911, at the annual meeting of the American Society of Mechanical Engineers. Generally accepted psychrometric data of the time were based on an empirical formula deduced from many simultaneous readings of dry-bulb, wet-bulb, and dew-point temperatures of air. Carrier questioned the accuracy of the data and stated that there were errors in the form of the equation as well as in the constants employed, which made the inaccuracy more pronounced at low and at high humidities. He proposed new psychrometric formulae based on principles reasoned out theoretically and then proved by practical demonstration. Carrier wrote:

> The following principles underlie the entire theory of the evaporative method of moisture determination, as well as of air conditioning:
>
> (A) When dry air is saturated adiabatically the temperature is reduced as the absolute humidity is increased, and the decrease of sensible heat is exactly equal to the simultaneous increase in latent heat due to evaporation.
>
> (B) As the moisture content of air is increased adiabatically the temperature is reduced simultaneously until the vapor pressure corresponds to the temperature, when no further heat metamorphosis is pos-

sible. This ultimate temperature may be termed the temperature of adiabatic saturation.

(C) When an insulated body of water is permitted to evaporate freely in the air, it assumes the temperature of adiabatic saturation of that air and is unaffected by convection; i.e., the true wet-bulb temperature of air is identical with its temperature of adiabatic saturation.

From these three fundamental principles there may be deducted a fourth:

(D) The true wet-bulb temperature of the air depends entirely on the total of the sensible and the latent heat in the air, and is independent of their relative proportions. In other words, the wet-bulb temperature of the air is constant, providing the total heat of the air is constant.

Carrier's paper was an all-important milestone in air conditioning. After its publication engineers accepted the control of air as a branch of their profession. Carrier's psychrometric chart was reproduced in college textbooks and engineering schools began to include air conditioning as a subject to be covered by their students. His "Formulae"—translated into many foreign languages—became the authoritative basis for all fundamental calculations in the air conditioning industry. It not only brought scientific recognition to Carrier, then only thirty-five years old, but also gave an impetus to the air conditioning industry headed by the company which bears his name. Sales by the Carrier Air Conditioning Company of America rose to sixty-three in 1912, to ninety-three in 1913, and to one hundred and thirty in 1914. They included air conditioning systems for paper and textile mills, malt houses, department stores, hotels, pharmaceutical plants, soap, rubber and tobacco factories, candy and processed food plants, film studios, breweries, bakeries, and meat packing houses.

Just as the Carrier Air Conditioning Company of America was beginning to prosper, Germany invaded Belgium and World War I was under way. This tragic event caused many American businesses to pause and take stock. Among these was Buffalo Forge, whose management had already begun to question the advisability of engineering and installing air conditioning systems, even indirectly.

Late in 1914 Buffalo Forge Company decided to confine its activities entirely to manufacturing. This important policy decision made necessary a great change in the business of the air conditioning subsidiary. William F. Wendt journeyed to New York to break the news to Irvine Lyle. On the day of his arrival, he found himself having breakfast in the same restaurant with Willis Carrier, who years later said this of the unexpected meeting:

> As soon as he saw me, Mr. Wendt asked that I join him at his table. While we ate, he told me what had been decided. And he said it looked like everyone in the air conditioning company would have to be let out except Irvine Lyle, who would be offered his old job as manager of the Buffalo Forge sales office in New York. I was not an employee of the subsidiary.
>
> Mr. Wendt suggested that we go together to see Lyle and then decided against it. At that point we both stopped talking. The breakfast ended in silence.

VII

WILLIS Carrier and Irvine Lyle were heartsick over the decision reached by the top management of Buffalo Forge Company. The fact that they would still have jobs was of no moment to them. What both men saw was that their long, hard work in building the air conditioning industry would now either sink into oblivion or be carried on by others. Neither prospect was tolerable. In New York, Lyle did not tell the engineering staff for a time, hoping against hope for a reversal of the verdict. Back in Buffalo, Carrier concentrated on the problem, finally decided the answer was to form a new company. "If Irvine Lyle is willing to take the chance," he reasoned, "and if we can get his staff of engineers to come along with us, we could make a go of it."

Carrier went to New York to talk it over with Lyle. The two asked Murphy to come up from Philadelphia, and the three went to Lyle's home to discuss the pros and cons of forming a new company. The result was that seven young engineers banded together, staked their fortunes and futures, grasped opportunity, and started the Carrier Engineering Corporation, incorporated in New York State on June 26, 1915. The seven engineers were Willis H. Carrier, J. Irvine Lyle, Edward T. Murphy, L. Logan Lewis, Ernest T. Lyle, Alfred E. Stacey, Jr., and Edmund P. Heckel. Their legal advisor was Carrier's friend and neighbor, Charles J. Staples, of the Buffalo law firm of Mitchell and Staples. For his fee, Staples agreed to take stock in the infant company and thereby in 1915 began an association which lasted until his death in 1949. For more than a third of a century he provided legal guidance, directing procedures when the company in 1930 merged with two others, and serving as a director and secretary of the Carrier Corporation that resulted from that merger.

In 1948 Staples, looking back on the seven young engineers —"the original seven"—who started the company, said:

> I never worked with a finer group of men. When I suggested naming the company the Carrier Engineering Corporation, and asked Mr. Lyle what he thought of it, he was quite satisfied. Irvine was never the jealous type.
>
> As the new company was to function as a sales and engineering organization, much as the former company had done, it needed connections with a manufacturer of air conditioning apparatus. Buffalo Forge Company was the logical manufacturer. By this time Willis and Irvine no longer resented Wendt's retrenchment program, and welcomed an arrangement whereby they could sell Buffalo Forge apparatus.

Carrier was president of the new company, Lyle treasurer and general manager. Under the guidance of Staples, they signed an agreement to hold an equal number of shares of stock and to have equal authority, each pledging never to attempt to usurp power from the other. Fifteen years passed without either finding any cause to refer to their agreement. Then, in 1930, when their company was merging with two others and stock was being reclassified, they decided to cancel their original agreement. No one could find a copy. In order to cancel their agreement, Carrier and Lyle, with Staples checking, wrote a joint letter setting forth the contents to the best of their memory.

When Carrier Engineering Corporation was formed in 1915,

Carrier and Lyle each agreed to give their services to the new company for six years, while the five other engineers agreed to remain with it for three years. There were various agreements with the Buffalo Forge Company and one with the Carrier Air Conditioning Company of America. This company retained the right to sell Carrier-designed air washers for ventilating and fan-heating systems, while Carrier Engineering Corporation concentrated on obtaining air conditioning and related contracts.

When all the agreements were reached and signed "the original seven" were still without any real money. They were good friends, they knew more about air conditioning than any other group, they were rich in courage and ingenuity, but they had no capital. They went ahead anyway, agreeing to start operations once stock subscriptions totaled $2,500. This meager total was reached by June 26, 1915, when the first stockholders meeting was held at the Iroquois Hotel in Buffalo. There were five stockholders, each of whom had subscribed to five shares of $100 each: Willis H. Carrier, J. Irvine Lyle, Edward T. Murphy, Henry W. Wendt, and Jennie Martin Carrier, whom Willis Carrier had married in 1913 following the death of his first wife a year earlier.

So on July 1, 1915, Carrier Engineering Corporation opened its offices, in New York, Chicago, Philadelphia, Boston, and Buffalo. The headquarters were in Buffalo, two rooms rented in the Mutual Life Building for use by Willis Carrier, staffed by a secretary and a draftsman. Years later the secretary said:

> We bought the minimum in furnishings and spent as little as possible for what we did buy. We ended up with second-hand furniture—two desks, a drafting board and stool, and a few files. We had two wicker chairs for visitors, and Mr. Carrier's friends would ask him if he had swiped them from a tavern.

To stretch the original $2,500, Carrier paid Murphy's salary out of his own pocket and Irvine Lyle paid Lewis's. All salaries were cut to what Murphy later termed "a subsistence level." Carrier made stock sales to friends, neighbors, and business acquaintances, including his dentist and the owner of a silk mill Carrier had air-conditioned. By August both Carrier and Lyle had purchased 38 shares, Ernest Lyle 18, Heckel 5. Murphy borrowed from a friend to buy 25 shares, while Lewis and Stacey each had to borrow $1,000 in order to buy 10 shares apiece. Altogether the original paid-in capital was $32,600. That was the total stock subscription on December 8, 1915, when the books were closed and no new capital was brought in until 1927, and then only for expansion purposes. By that year the capital and surplus of Carrier Engineering Corporation was $1,350,000, all built up from the original investment. In 1947 Murphy, considering what human elements and qualifications were present in this enterprising group, wrote:

> We had faith and enthusiasm in our enterprise, with loyalty to each other and to a common cause.
> We had courage and vision to seize opportunity when it appeared.
> We supplemented each other in all phases of an intricate business.
> We had a superior product, applied with sound engineering.
> We had a product that was needed in industry and took on the responsibility of leadership by continuous improvement in the product and in the broadening of the product's applications.
> We held steadfastly to a high standard of integrity in our products, in our engineering methods, and in our financial dealings with others.
> It would seem also that we had fate on our team. Perhaps it was because we had selected a business that contributes to better health and living for people by which a real service is rendered to mankind.

47

Eighteen days after Carrier Engineering Corporation opened its doors, Irvine Lyle closed the new company's first contract for $4,530. It was to air condition the American Munition Company's fuse-loading building at Paulsboro, New Jersey. The second contract was a $15,000 sale, to provide comfort air conditioning in Philadelphia's Masonic Temple. The third was a "fast-turnover" $950 sale, and the fourth was another munitions contract: installing a $32,000 air conditioning system for the International Arms and Fuse Company of Bloomfield, New Jersey. These first contracts differed from those of many companies then supplying mechanical equipment to contractors. Carrier Engineering Corporation did not bid on equipment specifications and did not guarantee apparatus by horsepower capacity or air volume. The company's commodity was "conditioned air" and it guaranteed its conditions.

By the end of 1915 more than forty contracts had been closed by the new company. They varied from large and small air conditioning systems to replacement parts, and even a school heating system. Carrier said: "We took every type of contract we could handle. We needed to bring in cash for our company."

Sales of air conditioning systems were made in the first six months to textile mills, tobacco factories, and bakeries as well as to munitions plants. Several air conditioning systems were installed to help the American Viscose Company extend its rayon production. Armour & Company ordered one for a new oleomargarine plant.

Meanwhile, Willis Carrier was exceedingly busy. He was spending time in the field, selling air conditioning apparatus, advising sales engineers, checking contract guarantees, devising a profit-sharing plan, and filing patent claims for two inventions—incidentally, the smallest number he had filed for many years. Carrier was also directing his attention to the development of an air conditioning system based on the ejector principle.

Carrier first observed the phenomenon of induced air flow in

1914 when he visited an air conditioning installation in a Holyoke, Massachusetts paper mill. He noticed that the air-distributing duct ran down the center of the paper drying room just under the ceiling, with outlets discharging the air to both sides. The paper to be dried was festooned on racks on each side of the center aisle. At first, Carrier thought that such air distribution was not good for the drying process, but when he found it worked satisfactorily he investigated to find out why. Then he made a discovery. "I was the one-eyed man," he said, "who saw something to which others had been blind."

What Carrier discovered was that the air, discharged horizontally across the ceiling, induced a secondary flow by pulling air upward from a clear space in the aisle under the duct. He began asking himself "How much and at what velocity must I have an air stream to 'pull in' enough air to double or triple the circulation?" He immediately made preliminary calculations, formed a mental picture with known data, and planned tests to obtain the missing data. Indeed, the prospects of an air conditioning system based on the ejector principle had been one factor that had encouraged Carrier and Lyle to start the new company.

Carrier figured he could condition about one-third of the volume of air employed in conventional distribution, introduce it into the room at high velocity, and thereby pull in sufficient "secondary air" to obtain the total volume needed to hold the temperature and relative humidity at the specified level. Such a system, he and Lyle knew, would employ smaller apparatus, lower costs, and widen markets.

However, when Carrier finished his design drawings and was ready to build experimental apparatus, Carrier Engineering Corporation had neither a laboratory nor funds to spend on research. To surmount this obstacle, Carrier formulated a unique and daring plan. He interested a manufacturer of enamelware in trying out the ejector system in his plant, sold him the apparatus, and thus turned the customer's factory into his own laboratory.

49

The manufacturer, the Republic Metalware Company of Buffalo, strove to get a mottled effect which was produced by a cycle of 15 to 30 minutes of slow drying followed by a period of rapid drying. Carrier got the manufacturer to build two tunnels—each 180 feet long—installed his ejector system to heat and circulate the air in the tunnels, and thereby obtained apparatus for tests on induced air flow. He guaranteed to dry 1400 pieces of enamelware an hour, ran into difficulties because of the variations in the sizes and shapes of the ware, but finally overcame them after tests and adjustments.

Negotiations with Republic Metalware were under way by December 14, 1915, and Carrier's unique contract was signed three months later. The experiments moved so swiftly that, by May 27, 1916, Stacey had sold a $3,217 ejector system to a Detroit manufacturer to dry 19,000 pounds of rubber in twenty-four hours. He guaranteed to remove 2.18 pounds of moisture a minute—sixteen gallons per hour. Stacey later said:

> Sixteen gallons of water per hour is a lot but I ended up more than doubling the amount. When starting up the system, all went fine—so good, in fact, the operating engineer kept adding rubber. I was able to keep up with him without much difficulty. When he reached sixty thousand pounds of rubber in twenty-four hours, the system still did the job.

Carrier and Lyle were now convinced that the ejector system had great possibilities. With proof of lower costs, speed-up in the drying processes, and controlled drying rates, sales efforts were intensified. They first interested macaroni manufacturers who were at that time taking forty-eight hours to dry each batch. Years later one manufacturer wrote:

> We simply had to dry more products in an established space, which Mr. Carrier guaranteed to do. He accom-

plished only half as much as he guaranteed, but he cut his bill in half, showing high moral principles. The procedure which he advocated has been somewhat modified; however, it is the basis for drying, used not only by us but by the entire macaroni industry.

The ejector system moved on into many other industries—fire brick, terra cotta, plaster molds, tiles, leather and hides, tobacco, rubber, candy, and ammonium nitrate. Carrier went back to the paper mills, whence had come his original idea, and designed a special system for drying paper. He spent much time experimenting in these mills, particularly one in Bangor, Maine. Some idea of how he worked was provided years later by two of his engineers who installed an air conditioning system in a macaroni plant. The macaroni, instead of drying properly, fell to the floor in millions of small pieces. The engineers inspected the system, found nothing wrong, filled two drying rooms, and started up the apparatus again. Twelve hours later they found 10,000 pounds of macaroni on the floor and telegraphed Carrier for help. One of them said:

> The Chief got to the plant in the late afternoon and looked over the job. We went to the hotel for dinner, where the Chief got cleaned up from his long trip. Then we went back to the plant, presumably to stay only a short time. We sat around a desk, the Chief in deep thought, and no one said a word. Directly, the Chief would get up, walk fast down the corridor and we followed him. Back at the desk we'd all sit down again, but after a few minutes the Chief would jump up, walk to the other end of the corridor, and we followed again. This went on all night, and hardly a word was spoken the entire time. By sunrise, the Chief had worked out a plan of action.

The Chief's plan was a series of tests based on length of drying periods. He started with 48 hours' drying and kept shortening the time until reaching the minimum at which the macaroni dried satisfactorily. We ruined a lot of macaroni before arriving at the minimum and safe drying period, and we paid for the ruined macaroni at the rate of five cents a pound. We would work the full time of the test, say fifty hours or more, sleep through a day and night, then start another run.

One "trouble job" with historic repercussions came after the sale in 1915 of two air conditioning systems for a time-fuse loading plant in Providence, Rhode Island. This plant was a joint operation of Westinghouse and American Locomotive, under the name of the Walco Company. When the Walco air conditioning system delivered only half its rated cooling capacity, a wire was dispatched to Carrier in Buffalo. He arrived at the plant, found there were three large ammonia reciprocating refrigerating machines—"certainly enough refrigeration to deliver the total rated capacity of almost 200 tons, but only producing half that amount of cooling."

To Carrier the cause of the trouble was obvious: the operating engineer, an "old-timer" in refrigeration, was running the system as though he were making ice instead of cooling water for air conditioning. All the coils were iced at the bottom. The back-pressure gauge read fifteen pounds. Carrier started to run up the back pressure and to adjust valves so that each coil would do its full quota of cooling. The operating engineer, fearing that the pipes would burst, ran out of the building saying, "I wash my hands of any responsibility of the damage you cause."

After reaching a back pressure of thirty-seven pounds and tediously adjusting all the valves, Carrier had the entire system operating as designed. He also had a new problem: to design a simple refrigerating machine. Later he stated that following the "trouble-shooting" at the Walco fuse-loading plant, his thoughts went along this line:

If air conditioning is going to grow, there will have to be something done about mechanical refrigeration. We must have a refrigerating system that is simple and foolproof—so simple that we can run warm water through a pipe and have it come out cold, just as we can now put in cold water and have it come out hot as in a boiler.

VIII

WHEN Willis Carrier first recognized the inadequacy of existing refrigerating machines, he was not able to tackle the problem immediately. He carried it with him several years, thinking about it on trains and at meals, once talking about it for an hour or more while standing in the rain with Murphy on a New York City street corner. As was characteristic of him, he had seen a need and evaluated what its solution would mean to engineers, to industry, and to business. Had his assay been less promising, he would have discarded the idea and proceeded no further with it. He described this pattern of action with these words: "I fish only for edible fish, and hunt only for edible game—even in the laboratory."

Carrier also had the characteristic of undertaking so many projects that he had insufficient time for all of them. But though he often postponed action on a problem, he never abandoned one that promised to be worth-while. A psychologist might argue that his carrying a problem around in his head for years, letting it simmer, may actually have facilitated his discovering a solution.

In any event, some time after sensing the need for an improved refrigerating machine he tried to interest existing manufacturers in improving their apparatus. When he informed one of these that the Baudelot coil was the most "unmechanical" piece of equipment in the entire system, the answer was, "It's the best we can do." Carrier said, "Some day I'll find a means to cool water as simply as we now heat it." Two years later, in June of 1920, Carrier had his thoughts on paper, in an eleven-page memorandum which Lyle passed on to the company's senior engineers with a most unusual injunction: "Please see that no one but you reads the attached discussion, and return it to me as soon as you have read it." The memorandum, titled "Development Possibilities for Improvement in Refrigeration," told of plans for a new type of machine. Carrier wrote:

> The entire system of electric transmission has been developed from nothing to an enormous industry with relatively simple motors that are high-speed rotative equipment. Industry has gone from low-speed reciprocating steam engines to high-speed rotative turbines. Pumping machinery is rapidly changing from reciprocating types to high-speed rotative pumps for both liquids and gases. Modern power plants have installed high-speed, direct-connected, centrifugal, boiler-feed pumps almost exclusively in replacing the old type of steam-driven reciprocating machines.
>
> Refrigeration, though classed among the older mechanical arts, has shown no such material progress. The same improvements that have taken place in electrical transmission and in steam machines and pumps must come in refrigerating machines.

Carrier then visualized the improvement to come: a refrigerating machine with a centrifugal compressor and direct drive suitable for the same high operating speed; the heat exchangers —compact, simple, effective, and low cost in comparison with the cumbersome equipment then used; and new refrigerants— non-toxic and having characteristics adaptable for operation with radically improved mechanical equipment. The memorandum of 1920 was a perfect example of Carrier's lifelong practice of analyzing a problem before taking action, building experimental equipment, and conducting tests.

In this preliminary stage Carrier and Lyle were compelled to consider a number of questions: whether the machine, when invented, should be built by others for their company to sell, and if so, how the company would then fare in the highly competitive refrigerating market—or whether their company should build the machine itself despite lack of experience in manufacturing. Carrier said:

> No matter what answer we reached, we were taking a chance on losing all we had built up over the years. But Lyle was willing to take the chance. We decided not to answer the question definitely until designs were better formulated. I marvel at the faith Irvine had in me even when my ideas approached the revolutionary in engineering thinking at the time.

To obtain experimental equipment with the least expenditure, Carrier considered existing rotary pumps and centrifugal compressors. Meanwhile he studied refrigerants, as the success of one depended upon the selection of the other. He later said:

> All I knew at the time was that I wanted a liquid with a high boiling point, around 110 degrees F, so I'd not have to employ high pressures or excessive vacuums in compressing the gas. I also knew I wanted a gas with as high a molecular weight as I could find, to keep the number of stages of compression to a minimum.
>
> The refrigerants in general use were not practical for my purpose. Ammonia would require many stages for centrifugal compression, and carbon dioxide more than I wanted to use. Pressure requirements for ammonia and carbon dioxide also ruled them out of my considerations. I had no choice but to find something new.

Carrier began combing a chemical dictionary for chemicals that met his specifications. He discarded sulphur dioxide, since disaster could follow if leaks occurred. He considered carbon tetrachloride but decided against it because it attacked metals, especially in the presence of water. In obtaining additional data on carbon tetrachloride, Carrier turned to a chemical textbook—presumably "General and Industrial Inorganic Chemistry" by Dr. Ettore Molinari—and there learned about dielene—$C_2H_2Cl_2$. He decided this was it, with the final decision contingent, of course, on exact knowledge of its chemical characteristics and physical properties. Dielene was not manufactured in the United States but the textbook mentioned a firm in Geneva as a source of supply. Carrier said:

> It looked as if dielene might be hard to get, but we determined to try it. When I wrote to the firm in Switzerland, the company informed me that dielene was manufactured in Germany and in industrial quantities, as it was used there as a cleaning fluid. Being manufactured in commercial quantities was certainly an advantage to us for it should be less costly than if produced for laboratory purposes only.

With the refrigerant tentatively selected, Carrier drew up specifications for the centrifugal compressor based on a capacity rating of fifty tons—a ton of refrigeration producing a cooling effect equivalent to the melting of a ton of ice during twenty-four hours. He then asked two American manufacturers to quote prices on building the unit. One quoted a price so high it would forbid competition with reciprocating machines then on the market. The other was not interested at all. Carrier said:

> I felt, as Carrier Engineering Corporation was basically an engineering and contracting firm, we did not have the necessary experience to become manufacturers, especially of heavy machinery where each part represented big money

to us at that time. But, when we could not interest a refrigeration manufacturer in building our machines, we had to build them ourselves. That was what Irvine Lyle wanted all the time. He felt we could build them—and with a profit.

Lyle began looking around for a factory with a modest size manufacturing area and sufficient office space to house the company's headquarters. Late in 1920 he found such a building at 750 Frelinghuysen Avenue, Newark, New Jersey. It was what he wanted, the price was right, and Lyle could have closed the deal immediately. Instead, Lyle followed a course which Carrier later described as follows:

> When Irvine Lyle and I began our talks about centrifugal refrigerating machines, our engineering force totaled about thirty. We were like a closely knit family, and Irvine was like a father to most of the men. All of their personal problems, financial and otherwise, gravitated to his office. He always lent a sympathetic ear, and he would try to straighten things out for them. The close relationships worked both ways. When we wondered whether to expand our activities—that is, go into manufacturing centrifugal refrigerating machines—Lyle said, "Let's ask the engineers."
>
> We had our annual meeting at the Old Fort Comfort Inn, Piedmont-on-the-Hudson, that year. After a dinner, Lyle told the men about our chance to buy a plant in Newark, of the amount of money involved—which was a lot for us, a small company—and of the risks we would be taking. He then called for comments. Everyone had a lot to say. All angles were discussed with vehemence. No one was timid about speaking up. When a vote was taken, it was overwhelmingly in favor of buying the plant.

The annual meeting took place early in January of 1921. The plant was purchased on March 10 and was occupied in May. Meanwhile in February, March, and April, patent claims were filed on the centrifugal refrigerating system which were finally issued in 1926: patents Nos. 1,575,817, 1,575,818, and 1,575,819. Carrier could not secure a basic patent on centrifugal compression of refrigerants because in 1913 Maurice Leblanc, a French engineer, had invented a four-stage compression machine using water vapor as the refrigerant—a machine which never got beyond the experimental stage. Carrier later explained:

> Since I could not get a basic patent, I decided to file claims for inventing a refrigerating machine using centrifugal compression. I tried to anticipate every possible patentable feature and cover them all in my claims as the inventor of the machine.

Carrier went to Europe in the late spring of 1921, ostensibly for pleasure, actually to get a manufacturer to build a centrifugal compressor to his specifications. When he called on one company in Switzerland and told the engineers what he wanted, they became so interested—"the type that would pick your brains"—that he picked up his papers and left in a hurry. He had a second Swiss company on his list but he figured their engineers might also be curious. So he went on to Germany and there arranged for the Leipzig firm of C. H. Jaeger & Company to build a compressor at less than one-sixth the price quoted by the American manufacturer. In Germany Carrier also located a dielene manufacturer and ordered two drums. In England he helped complete the organization of Carrier Engineering Company, Ltd., of England, and then returned to America. Here he found his company settled in new quarters, planning to expand into manufacturing, with everything staked on the Carrier centrifugal refrigerating machine, although it was not far beyond the drawing-board stage.

Tests, calculations, and then the design and construction of the condenser and cooler followed. The compressor arrived from Germany early in 1922. The completely assembled machine was put under tests with run-ins, alterations, more run-ins, and still more alterations. Carrier and Lyle became so confident that they set May 22, 1922, as the date for unveiling the machine. They invited three hundred engineers to have dinner at the plant and witness the ceremony. As an added attraction, they scheduled two boxing matches by local amateurs. Years later Carrier gave the following account of that historic event:

> When the day of the unveiling arrived, we had turned the machine over but we had not run it under load conditions. We did not have steam in our factory for the turbine, so we borrowed it from our next-door neighbor. By the time the steam reached us it was sufficient to run the machine when idling, but not to run it with the cooling load. By noon we had checked the steam line and fixed a leak. We started up the machine, using it to chill the water for our air conditioning system. When the guests arrived, our offices were cool and comfortable—and it was a hot day.

Following dinner in the center of the sheet-metal shop, the guests were addressed by Carrier on the development of the machine. He described the principles involved, and was explaining that the guests would later see the machine itself in the adjoining room, when a loud and continuous noise began coming from that room.

> It was terrible when I heard that long, loud, rumbling, slowly diminishing b-r-r-r-r. I visualized the rotor of the compressor tearing itself to pieces. Beads of perspiration came out on my forehead and my hands were soaking wet. But I kept right on talking, trying to act as if nothing had happened. Irvine, sitting near the back, casually left the

room with an air of calmness I knew he did not feel. Directly, he came back and signaled to me that all was okay. Later he told me the cause of the noise. In arranging the space for the boxing matches, one of our men pulled a large metal dining table across a rough concrete floor. No sound effects man could have done any better in imitating the disintegration of a rotative machine.

Thus was introduced the first major advance in mechanical refrigeration since David Boyle designed the original ammonia compressor in 1872. On March 26, 1923 the first sale of centrifugal refrigerating machines was made to Stephen F. Whitman and Sons of Philadelphia. This candy manufacturer bought three of them. A month later the second sale was made—to Wm. F. Schrafft and Sons of Boston. By the end of 1924 nine machines had been sold, including the experimental apparatus which was purchased by the Onondaga Pottery Company of Syracuse, New York. Twenty-eight years later this first machine is still serving a commercial purpose, cooling 238 gallons of water a minute to condition air in the pottery's lithographing plant.

When each machine was installed, Carrier operated it as test apparatus during the run-in periods. Each test brought modifications and changes, including more reliable means for removing air from the system and more complete elimination of moisture and foreign elements from the inner chambers. By the end of 1924 he decided a major change was needed in the design of the seal—the mechanism which, while preventing the refrigerant from leaking out where the drive shaft enters the unit, still permits free well-lubricated rotation of the shaft.

Carrier sent the German manufacturers instructions to separate the thrust and seal and suggested how to do it. When it came time for factory tests, he went to Germany to try out the new idea, but the seal did not operate as he had hoped. There in

Leipzig in early 1925, he worked day and night until one evening in his hotel room it came to him. He said:

> It was so simple I didn't know why I had not thought of it sooner. I put a valve, a sort of "doodad," in the oil line so that, as the compressor started up and the rotor moved axially toward the thrust end, the valve opened a little to let the oil through—a little oil at first and, as the speed picked up, the valve opened more, letting more oil through.

The "doodad" was a "restriction disc," thereafter installed on all compressors built to use dielene, and was not changed until a new refrigerant, Carrene 1, was adopted for the centrifugal machines.

Meanwhile, working with Lawrence C. Soule, Carrier had helped perfect lightweight finned coils, which presented a larger surface for heating, cooling, or dehumidifying air.

> Irvine Lyle worked out a wonderful plan. He reasoned that, if we built the coils for our air conditioning contracts exclusively, the sales would be much less than if we built them for the entire fan-heating industry. He proposed forming a sales organization made up of fan companies. As a result, Aerofin Corporation was founded in 1923. The company sells finned-tube coils. Our company manufactures them. Irvine's plan was a stroke of business genius and everybody concerned profited, including the customer.

IX

THE Carrier centrifugal refrigerating machine opened vast new fields to air conditioning. Prior to its introduction industrial plants comprised air conditioning's main market; most installations were made to condition air for products and processes, not for persons. The first sales of centrifugal machines to candy manufacturers and the like helped the company financially and enabled its engineers to develop refinements and improvements. But the sales represented no entry into the practically untouched market which Carrier, Lyle, and associates had visualized—comfort air conditioning.

The door to this market was partially opened in 1924 when Irvine Lyle made the first "comfort job" sale of the centrifugal refrigerating system to the J. L. Hudson Company department store in Detroit. This store had found that on bargain days the ventilating system in its basement was of little help. Temperatures soared; customers fainted. The manager feared he would have to discontinue bargain sales unless he could keep temperatures down when the crowds poured in. Lyle persuaded him to install an air conditioning system using three centrifugal machines, each of 195 tons capacity. The result was a cool basement, increased sales, and in a few years the extension of air conditioning to other floors. This was the first air conditioning installation in the department store field and it began a relationship which has continued through the years.

The real opening of the "comfort" market came, however, when centrifugal systems were introduced into motion-picture theaters. Carrier later said:

> Movies closed during hot weather or showed to such small audiences that they operated at a loss. Even on cool days the inside of the theater was hot if there were many people in the audience. The heat from the people was enormous. A ventilating system did not help much. We argued that, with air conditioning, the theater need never be dark and would do a box-office business in summer because people would go there to cool off—to be comfortable. With our air conditioning, we believed we could sell the theater market without much resistance.

So, with Carrier and Lyle encouraging them, the company's sales engineers began concentrating on theater owners. The engineers not only had Carrier's new, safe, and simple refrigerating system, but also a method for introducing cleaned, cooled, and dehumidified air into a theater, without causing drafts or cold feet. This no-draft feature was achieved through a system of by-pass down-draft air distribution which Logan Lewis had designed in 1922 for Grauman's Metropolitan Theater in Los Angeles. Designed, sold, and installed before Carrier's centrifugal refrigerating machine was available, this installation used a carbon dioxide refrigerating system to cool the air. The air flowed through outlets located in the ceiling, diffused slowly downward, then entered return grilles located in the floor. Thus, Lewis accomplished the seemingly impossible feat of circulating a large volume of cooled and dehumidified air without the audience being aware of any air movement. This has caused many persons to refer to Grauman's Metropolitan Theater as "the birthplace of theater air conditioning." However, the crucial test of theater air conditioning, and one of the most decisive moments in all air conditioning history, was to come at the New York Rivoli Theater in 1925.

This test was preceded and to some degree produced by air conditioning sales to three Texas theaters owned by Will Horwitz, Jr.—the Palace in Dallas and the Texan and Iris in Houston. In the Palace Theater between June and August of 1924, Carrier's centrifugal refrigerating apparatus was installed for the first time in any theater, combined with Lewis's by-pass downdraft system. The theater advertised "cool and clear" weather and a consulting engineer on the job later reported that "as proof that it was easy to operate, the man who had previously been a sort of janitor was taught to run this unit and did it well."

In Houston the two Horwitz theaters were cooled by one large centrifugal machine, located in the Texan. Chilled water was pumped to a storage tank in the Iris across the street. Horwitz wrote:

> The cooling plant is revolutionizing picture show attendance in Houston. Each patron exclaims with delight when he gets inside the doorway. The plant is working perfectly. Our engineer says he has nothing to do on the job but loaf.

Word of the Texas installations spread. The Rivoli in New York decided to discard its ventilating system and install air conditioning. On November 20, 1924, the Rivoli contracted for a 133-ton machine and the Carrier Engineering Corporation was faced with a critical test. The company's whole future in theater air conditioning was at stake. To be sure nothing was overlooked that would affect results, Carrier himself and the company's top engineers all worked on the job, drawing up layouts, checking, and re-checking—aiming at the finest comfort air conditioning system that could be designed.

Meanwhile, it developed that the New York City Building Code barred this type of installation because dielene was not listed as an approved refrigerant. In fact, it was not listed at all. The Carrier engineers undertook to get a permit. They called on the safety chief, told him of fifteen machines then in operation, all with no-accident records, showed him reports from independent consulting chemists and engineers who had run three types of tests on dielene. The safety chief was unconvinced. Later Carrier said:

> I then tried an experiment. Right in his office I poured some dielene into a container and dropped a lighted match in it. Well!—the safety chief got mad—and scared, too, I think. He said if we were going to try such stunts, we would have to go elsewhere. All the while the dielene burned downward very slowly, no flare-up, no explosion. Finally he was sufficiently convinced that dielene was safe, and granted the permit with the stipulation to isolate the machine and take many precautions beyond the code requirements.

The Rivoli was scheduled to open on Memorial Day in 1925. Partly because of the delay in getting a permit, there was difficulty in meeting the deadline. Carrier and three other top men stayed up practically the whole night of May 29 seeing that everything was set up and ready to go. An unsigned, undated memorandum in Carrier's files, presumably written by him, records:

> Typical of show business, the opening of the Rivoli was widely advertised and its air conditioning system heralded along Broadway. Long before the doors opened, people lined up at the box office—curious about "cool comfort" as offered by the managers. It was like a World Series crowd waiting for bleacher seats. They were not only curious, but skeptical—all of the women and some of the men had fans —a standard accessory of that day. . . .
> Among the spectators was Adolph Zukor. I recall dis-

tinctly how quiet and reserved he was when he walked in and took a seat in the balcony. Zukor may have come from California, but he was there to be shown!

Final adjustments delayed us in starting up the machine, so that the doors opened before the air conditioning system was turned on. The people poured in, filled all the seats, and stood seven deep in the back of the theater. We had more than we had bargained for and were plenty worried. From the wings we watched in dismay as two thousand fans fluttered. We felt that Mr. Zukor was watching the people instead of the picture—and saw all those waving fans!

It takes time to pull down the temperature in a quickly filled theater on a hot day, and a still longer time for a packed house. Gradually, almost imperceptibly, the fans dropped into laps as the effects of the air conditioning system became evident. Only a few chronic fanners persisted, but soon they, too, ceased fanning. We had stopped them "cold" and breathed a great sigh of relief.

We then went into the lobby and waited for Mr. Zukor to come downstairs. When he saw us, he did not wait for us to ask his opinion. He said tersely, "Yes, the people are going to like it." That was a jubilant moment for us—we had passed the "acid test."

On July 3, 1925, the managing director of the Rivoli wrote that, "Although the apparatus has only been in operation four weeks, it is the talk of Broadway." Subsequently, the architects for the new Paramount in New York, who had specified an up-draft ventilating system, were persuaded that their designs were obsolete even though the drawings were still on the drafting board. They discarded the up-draft for by-pass down-draft air distribution and specified air conditioning in place of ventilation. The $91,148 Paramount system, sold January 25, 1926, was the sixteenth theater air conditioning contract signed by Carrier Engineering Corporation. While the Paramount was under construction, Roxy—the noted theater promoter—began negotiations to build America's largest motion-picture house. Carrier later said:

> Roxy's backers, familiar with the year around capacity attendance of the Rivoli, agreed to finance the project only if an air conditioning system designed by us, or its equivalent, be installed in the theater. Ed Heckel, then in our Chicago office, persuaded the architect to specify our equipment. After the plans were completed, the fight for the contract was transferred to New York City. It was a fight, too, in spite of our favorable position. Irvine Lyle and his brother, Ernest, plus Ed Heckel's preliminary efforts, finally landed the job for us. It was a big one—two centrifugal refrigerating machines, each with 210 tons capacity. By that time we had changed our refrigerant from dielene to methylene chloride—Carrene 1. The authorities no longer questioned the safety of our equipment. However, it did take a lot of engineering, arguing, and planning before we filled the command performance to air condition the Roxy, a six-thousand-seat theater.

Though theater sales were exceeding their expectations, Carrier and Lyle did not rest on their achievements. The company formed a theater sales organization to concentrate on this market. Irvine Lyle spent much time in Manhattan helping to close contracts. Willis Carrier advised on the unusual engineering problems which arose with each prospect, directed the installations, and supervised try-out runs and acceptance tests. The success at the Rivoli started a demand which spread throughout the motion-picture world. By 1930 the company had air conditioned some 300 theaters. Some that had slumped or even closed during hot weather now reported attendance going up in both July and

August. The motion-picture theater became the summer stay-at-home's sea breeze. Meanwhile, newsreel theaters were starting up and they, as well as the small neighborhood houses, needed air conditioning to attract and hold patrons in order to meet the competition of the large air conditioned downtown theaters.

But many small theaters did not have ceiling room for air-distributing ducts. For them, Carrier developed a horizontal discharge system. He applied ejector nozzles, long used in dryers, to theater air conditioning. Placed near the ceiling in the end wall, the nozzles discharged the conditioned air horizontally toward the front of the theater at a high velocity. As in the dryers, the stream set up an induced air flow so that the theater air mixed with the conditioned air, and the dry, cool mixture kept the audience comfortable. Carrier's ejector system of distribution for theaters, for which patent claims were filed in 1927, not only eliminated the need for ceiling room but also reduced initial costs.

Meanwhile in the 1920's, Willis Carrier was busy with a variety of projects which made Carrier Engineering Corporation first and foremost in air conditioning. In 1925 he and Irvine Lyle, finding that a new Madison Square Garden was to be built in New York and that an ice-skating rink was included in the plans, decided to try to close the contract. They saw that here was a chance to disprove competitors' statements that the centrifugal refrigerating machine could not do more than chill water for air conditioning systems. The promoters of the new, actually the third, Madison Square Garden were George L. Rickard, known as "Tex," and John Ringling, whose reputation as a showman rivaled that of P. T. Barnum, builder of the first Garden. That first building was roofless. The second had a sliding glass roof, and it was here that the long 1924 National Democratic Convention sweltered for days. Rickard and Ringling hoped to create the illusion of a garden, not with a roofless building or sliding

69

roof but by conditioning the air. In 1948 Carrier recalled the negotiations with Rickard and Ringling:

> We did not have much trouble convincing them that we should have the contract for the air conditioning system. It was a different matter when it came to using our refrigerating machine to freeze the ice for the skating rink. Tex had heard the disparaging remarks made by our competitors about our centrifugals not being applicable for low temperatures. He and his associates came out to our plant to check for themselves—to see our experimental machine making ice. We did not have it down to freezing when the visitors arrived, so I took them into the office for a conference while we continued to try to bring down the temperature. We found the steam ejector purge had failed, fixed it, and were freezing ice by the time the conference ended.

The demonstration impressed but did not fully convince Rickard. However, Irvine Lyle, the master salesman, on July 21, 1925, closed the contract for three refrigerating machines, totaling 800 tons capacity, to cool the air in summer and freeze the ice on the 110- by 180-foot skating rink in other seasons. Carrier worked closely with the erection men to forestall trouble on the first freezing application of his centrifugal machine. He later said:

> Something did go wrong, but through no fault of ours. However, it did cause us a bad night and we barely squeaked through. We finished installing the centrifugal machines ahead of the plumbers, and the machines stood idle for about a month waiting for their piping connections. When we got the chance to run the machines, our brine pump did not work. Our erection superintendent reasoned that, as we were circulating such a high concentrate brine, it would not freeze, and thermometers in the line indicated

we were at the right temperature levels. The trouble must be in the pump.

But I saw frost on the coils in the evaporator and that meant a freeze-up, regardless of other signs. Maybe the brine was not well mixed, maybe the thermometers were too short and not down in the liquid. Frost meant freeze-up. During the long wait for plumbing connections, water had collected in pockets in the refrigerating circuit, and when we turned on one of the machines the water froze.

We turned off the machine and let it warm up. We then drained out the refrigerant and ran it through calcium chloride to get the water out. If the mixture of air and refrigerant had been just right, we would have got an explosion—but nothing happened. We put the dehydrated refrigerant back in the machine and all was okay.

This freeze-up occurred just six days before the Madison Square Garden ice arena was scheduled to open with a hockey match between the New York Hockey Team and Les Canadiens of Montreal. Draining the machines and dehydrating the refrigerant was a long process. As time passed, tension tightened. Stanley L. Groom, managing director of Carrier Engineering Company, Ltd., London, England, was visiting in America at the time and in 1951 recalled the scene at the Garden:

> As the day of the opening approached, everybody concerned was almost to the hair-tearing stage. Tex Rickard walked up and down, as worried as any person I'd ever seen. Just when the atmosphere was charged with anxiety, in walked the Chief. He carried his ice skates.

Another significant contract in the 1920's was to air condition the Chambers of the Senate and House of Representatives in the U. S. Capitol. Carrier and Lyle had figured on a system for the Capitol in 1921, but nothing came of it because they did not have a suitable refrigerating machine. Now, with the centrifugal and by-pass down-draft air distribution, they thought they would try for the contract again. Carrier later said:

> Irvine Lyle planned the strategy. We needed strategy to sell air conditioning where there would be no tangible returns, as increased retail sales or filled theaters. The cost of air conditioning the Senate and House would be considerable, and the members voting on the appropriations, being laymen, would not understand what they were paying for. To have men informed on air conditioning pass on the proposal was part of Lyle's strategy.
>
> Lyle visited Mr. David Lynn, the architect for the Capitol building, and presented an estimate for the work which he had computed at our office. We made the amount sufficiently large to avoid upping the appropriation if the proposition reached a bidding stage and we'd missed something in our rough estimate. Lyle recommended that the government select a committee of experts in the heating, ventilating, and air conditioning field to advise the action to be taken on the idea.

After Lyle got Lynn to name an advisory committee, made up of experts, Carrier and Stacey went to work in the laboratory. Air conditioning the House posed one particularly difficult problem because the ceiling was made up of large glass panels and the architect had stipulated "no change in the interior." Hunting for a solution of this problem, Carrier and Logan Lewis decided to raise each glass panel about two inches and insert long narrow outlets in the risers, for the introduction of conditioned air. Carrier said:

> Nothing like this had ever been done before and we did not know whether we could put in all the air we needed without causing drafts. We had to run tests to be sure.

Stacey built in our laboratory what today we'd call a "mock-up"—one ceiling panel and outlets in the frame underneath, all to full scale. We blew air through the outlets at the rate and temperature we'd figured for the actual installation. We tested for drafts with lighted candles and found we could introduce the air without a flicker of the flame at head levels. The most sensitive congressman with the baldest head would have no cause to complain of drafts. We took pictures of the tests.

Carrier Engineering Corporation got the contract, although its bid was the highest of the three submitted. The advisory committee, which made the recommendation, stressed the economy, safety, and simplicity of operation with the Carrier-designed system. One committeeman cited the success of the Carrier systems in the Rivoli, Paramount, and Roxy theaters and showed photographs of the laboratory tests on air flow. The House system went in operation in December of 1928 and the Senate system in August of the following year.

Meanwhile, Carrier had designed an air conditioning system for the Morro Velho Gold Mine in Brazil. The operators believed that, with conditioned air, they could reduce the intense heat experienced 7,000 feet below the earth's surface, where rock temperatures measured 130 degrees F. Hoping that cooler air would improve working conditions, increase production, and enable diggings to be extended to even lower levels, they had consulted Carrier in 1923. After correcting readings of his psychrometric chart from sea-level pressures to 1.2 atmospheres, he completed his first design for a mine air conditioning system. Carrier advocated placing the air conditioning apparatus and the refrigerating machine at working depth. His arrangements thus differed materially from ventilating practices employed in mines, where the fans were located at the surface.

The operators accepted Carrier's recommendations, but Irvine

Lyle advised the installation be postponed. He felt that more field experience was desirable before a refrigerating machine in production less than a year should be installed so far from home. Lyle, always practical, foresaw that trouble might involve traveling expenses exceeding profits from the sale.

Carrier applied for a patent on a "System for Cooling Mines and Other Chambers Requiring Ventilation" in 1924. It was issued in June of 1929; which, coincidentally, was the summer the company accepted a contract to condition 20,300 cubic feet of air per minute for the gold mine in Brazil.

Early in 1924 Carrier also spent some time adapting his centrifugal machine to chilling water for air conditioning the engine room of the U.S.S. *Arkansas,* where the heat rivaled that of the gold mine. As an extra precaution, he recommended the substitution of trichlorethylene ($CHClCCl_2$) for dielene. The Navy tested and approved the new refrigerant, Carrier began making adjustments, and the negotiations begun in 1924 were completed in 1925. By then orders came for Carrier Engineering to deliver the machine to the U.S.S. *Wyoming* rather than the U.S.S. *Arkansas.* So the *Wyoming,* with a special two-stage unit of 95 tons capacity, installed late in 1925, became the first ship equipped with a centrifugal refrigerating machine.

Carrier never used trichlorethylene again. With the advent of the "Freons," whose availability was made known in 1930 by Dr. Thomas Midgley, Jr., a standard refrigerant was available which met Navy specifications as well as opening the way to the development of economical air conditioning of small enclosures. Moreover, Carrier himself had soon after the installation on the *Wyoming* been put on the track of a new refrigerant for the centrifugal refrigerating machine: methylene chloride (CH_2Cl_2). He and Stacey conducted tests, calling in a consultant, and secured findings so favorable that methylene chloride was accepted as the refrigerant for future centrifugal machines. In 1926 Carrier filed claims for a patent for methylene chloride as a refrigerant.

Carrier Engineering Corporation named it "Carrene," which later became "Carrene 1." Carrier said:

> With Carrene 1, we doubled capacity of the machine obtained with dielene—the horsepower doubled also. By getting twice the refrigeration without increasing the machine size, we reduced the cost per ton of cooling and greatly improved our competitive position pricewise with reciprocating machines. We did not wait for a new-design machine for the new refrigerant, but began selling the current type with Carrene 1 in place of dielene. We did this to get the advantage of the increased capacity it afforded.

X

IN the early 1920's, after developing centrifugal refrigerating apparatus which introduced air conditioning into vast new markets, Willis Carrier began pioneering with small air conditioning units. Owners of small retail shops wanted air conditioning to compete with the large air conditioned department stores. To get into this market, Carrier Engineering Corporation designed a commercial apparatus called a "unit air conditioner," which performed all the functions of the central station system. The store owner who needed only 2,500 cubic feet of air per minute could now offer his customers the comfort of cooled air on hot summer days. Similarly, the small manufacturer had available to him air conditioning equipment equal to that sold to large plants.

The sale of these units appeared so promising that the company, in 1928, set up a Standard Products Division to cover the market. The first sale was made in February of 1928: a unit conditioner with a capacity of 2,500 cubic feet of air per minute to the Merchants Refrigerating Company for controlling the air in an egg storage room in its Newark, New Jersey, warehouse. More sales followed, especially to retail shops and industrial laboratories. Unitary air conditioning was on its way.

Also in 1928 the company began marketing a residential "Weathermaker." This was actually a winter air conditioner—heating, humidifying, cleaning, and circulating the air in the home during cold weather—with provision for adding summer cooling. Since cold water was the only source of cooling, Carrier and his associates planned the development of a refrigerating machine suitable for home application. To develop the residential market fully, they formed a subsidiary, known as the Carrier-Lyle Corporation, which operated until the early 1930's. Then, during the depression the market virtually disappeared.

The year 1928 also brought plant expansion. A second building was purchased in Newark, a block from the first. To distinguish between the two plants, the first was named the Carrier Plant and the second the Lyle Plant. When a third plant was erected the next year in Allentown, Pennsylvania, the Carrier Engineering Corporation had manufacturing floor area totaling 231,550 square feet. Carrier later said:

> From none in 1921 to over five acres in 1929 was quite a jump for us, but peanuts compared to our present plant. However, it proved our decision, sanctioned by our engineers when presented to them in a sales meeting, to enter the manufacturing end of our business had been a wise one.
>
> But the acquisition of the second building, the Lyle plant and offices, really started something. Irvine said he wanted us to install in it the best air conditioning system ever designed for an office building. There was no good system for multi-room buildings at that time, so we had to design one. For years I had thought that air should be introduced into the room at the window-sill level and sweep upward over the glass. In winter the heated air would cut down the chilling caused by the cold glass. In summer the cooled air would cut down the heating caused by the sun-heated glass. With our own office building, we could experiment with such a system.
>
> We installed a central station air conditioning apparatus with conventional ducts leading to the under-the-window connections. In each we placed a heating coil to add the

extra heat needed to warm the glass. It worked out fine—was our first step toward the Weathermasters that later made the air conditioning of skyscrapers practical.

In one office only did we apply the ejector principle of air flow which we had found so successful in dryers. The dryers had exposed nozzles. In the office installation we enclosed nozzles in a wall duct connected to an air inlet grille near the floor and an air outlet register near the ceiling. By discharging a small quantity of apparatus air at high velocity through the nozzles, we set up an induced flow which pulled air from the office through the lower opening. This secondary air mixed with the primary air. The mixture discharged through the upper opening. That was our first Weathermaster, and it was very effective. We knew we had the idea for an air conditioning system for multi-room buildings. However, many events kept us from perfecting the system for several years.

While all this was going on, Willis Carrier and his associates became interested in railroad air conditioning. Back in 1884 the Baltimore and Ohio Railroad had tried to cool a passenger car by passing air over a huge icebox built at the head end. For obvious reasons, this approach failed. In 1907, while designing a system to cool freight cars for the Santa Fe, Carrier proposed an apparatus room located below the station platform in which air would be cooled for delivery to the cars while they stood in the station. This system was never installed. In 1913 the Pullman Company, according to Carrier, "got interested in car cooling and asked us to design equipment, but we lacked suitable refrigerating machines at the time, so did nothing about it." In 1929 a representative of the B & O called on Carrier and asked him to design an air conditioning system for a passenger car, to be tested in the railroad yards at Baltimore. Carrier knew that refrigerating machines suitable for railroad car air conditioning were still in the development stage, but he set out to do the best he could. He considered evaporative cooling, discarded that approach, and recommended a five-ton ammonia machine. He then laid out the first railroad car air conditioning system, which consisted of one spray-type unit to cool and dehumidify the air, a similar unit to serve as a cooling tower for the condensing water, and an ammonia refrigerating machine. In the summer of 1929, this system was installed on the B & O coach No. 5275. Carrier said:

> We also began research on railroad car air conditioning in our plant in the summer of 1929. Stacey built a mock-up of a dining car to study air distribution, which was as necessary to the success of the system as proper refrigeration was to its safety and practicability.

Early in 1930, after running tests on coach No. 5275, the B & O ordered an air conditioning and refrigerating system for the diner, "Martha Washington," of "The Columbian," a crack train operating between New York and Washington. The system consisted of a 2,500-cubic-feet-per-minute, horizontal air conditioning unit placed overhead at the car entrance, two air-distributing ducts located above the ceiling with diffuser-type, down-discharge outlets, an ammonia reciprocating refrigerating machine mounted under the car, and a cooling tower in a locker. On April 14, 1930, the diner was tested on a run between Baltimore and Cumberland. To obtain summer conditions for the test the heating system was turned on, the diner heated to 93 degrees F, then the heat turned off and the air conditioning system turned on. In twenty minutes the temperature dropped to 73 degrees F. The first air conditioning system in a railroad car had succeeded in producing comfort.

That summer a special train, including the diner, was organized for the journey to the Minneapolis meeting of the American Society of Heating and Ventilating Engineers. Willis Carrier was invited to be a guest of honor at a dinner in the diner where he was hailed as "The Chief of the Air Conditioning Industry," and the B & O given recognition as a "pioneer railroad." The diner was shown at the 1930 convention of the American Railway Association in Atlantic City. Some 3,000 railroad engineers observed that, while temperature in adjacent cars read 96 degrees F, it was only 73 degrees F in the "Martha Washington." Later in 1930 the Santa Fe ordered a similar system for a diner, and the Missouri, Kansas & Texas Railroad ordered systems for three of its dining cars.

By this time Carrier was far along in the development of a new type of small-tonnage refrigerating machine—the steam ejector unit—designed to revolutionize railroad car air conditioning and sweep into many other fields. Carrier began his research on the steam ejector refrigerating principle early in 1929. He was attracted to the simplicity of using a steam jet to create a vacuum in a partially-filled water tank, causing the water to boil at a low temperature, and thereby cooling the water to 40 degrees F or lower. He saw the advantages of no moving parts, an unquestionably safe refrigerant, and utilization of existing steam plants. Many steam ejector systems were operated for chilling water for industrial processes, but they had not been applied to chilling water for air conditioning when Carrier, Stacey, and associates began their research.

On November 3, 1930, Irvine Lyle wrote to Stacey, who was director of research, "Mr. Carrier has approved the purchase of a railroad car for test purposes." Later Stacey stated:

> We bought an old day coach from the Central of New Jersey for $400, rolled it onto tracks in the yard at our plant at 750 Frelinghuysen. Then we built a shed over it and installed equipment to maintain in the shed summer operating conditions. We made it a hot summer in the shed, holding temperatures up to 106 degrees F dry-bulb and 84 degrees F wet-bulb. We simulated the effect of a moving

train by blowing air over it at various rates. We actually had a wind tunnel, but we did not give it such a name.

Inside the car a 100-watt electric light bulb was burned in each seat to produce the heat equivalent to that given off by passengers. Later Stacey built partitions and changed the coach to resemble a sleeper to study means for air distribution in Pullmans with the berths made up. Research engineers used the car for tests to determine the best possible air conditioning system for railroad cars. By May of 1931, a five-ton steam ejector refrigerating machine was designed, completed, installed, and more tests started. On August 21, 1931, the new system of railroad air conditioning was shown to several hundred railroad men. They heard Carrier discuss the apparatus technically, stepped into the shed, got a blast of the hot, dirty, dusty air, then entered the cool clean passenger car. They got the point. In a few months the Santa Fe was ordering air conditioning apparatus with steam ejector refrigeration for twelve cars. The Union Pacific in February of 1932 ordered the system for three cars. The Milwaukee put it in its crack train, "The Hiawatha." Other railroads followed. The result was not only a new market for air conditioning, but also an expansion of old markets.

Getting into the railroads was certainly one of the industry's greatest forward steps. It not only carried air conditioning into the nooks and crannies of the land but also exposed to it people with much influence. The movies brought it to the masses, the first department stores mainly to women. Air conditioning began spreading faster than ever when businessmen began stepping out of the heat into cool Pullmans and diners—businessmen who could order it for their own plants, shops, offices, and homes.

XI

IN 1930 Carrier Corporation was formed through the merger of Carrier Engineering Corporation with Brunswick-Kroeschell Company and York Heating & Ventilating Corporation. The two latter concerns were primarily manufacturers, while Carrier Engineering Corporation was both an engineering and manufacturing enterprise, although essentially the former. Brunswick-Kroeschell Company was an outgrowth of two boiler companies founded before the start of mechanical refrigeration in the United States. Its president was Sidney B. Carpender, a graduate engineer from Cornell University in 1907. Carpender, Carrier, and Irvine Lyle had often met at engineering society meetings, served on committees together, and become personal friends. In 1928, the three attended the spring meeting in Hot Springs, Virginia, of the Refrigerating Manufacturers Association. They played golf together and talked business during the game. Two decades later, Carpender wrote of their conversation on the golf course:

> Our company could see the handwriting on the wall with reference to the end of ammonia machine manufacturing for comfort cooling, and suspected something of the same with regard to carbon dioxide. We were a little worried about Carrier patents with regard to our business in air conditioning. However, of more vital importance to our

companies was the competition in air conditioning sales we gave each other. I thought we might pool our resources for our mutual benefit, instead of fighting each other. Carrier Engineering Corporation needed our carbon dioxide machine for certain types of work too small for the centrifugal machine.

Both companies lacked working capital for expansion, and we thought that we could secure that capital through a merger better than individually, especially because, in a merger, we could pool our engineering and sales efforts and effect great savings. We had two very valuable sales outlets which could be vastly increased by a merger: namely, the marine and export businesses. Conversations about our business preceded the one in Hot Springs, but it was on the golf course that I suggested the merger to Dr. Carrier and Mr. Lyle.

Carrier and Lyle favored the proposed merger, which promised the merchandising organization needed to sell refrigerating units as well as an enlarged staff of experienced engineers. So great seemed the potential advantages of merging with a refrigerating company that Carrier and Lyle extended the idea to include York Heating & Ventilating Corporation. This firm, organized in 1919, built and sold unit heaters, possessed a merchandising organization, and had made some progress in developing a room-cooling unit. Its president was Thornton Lewis, who was a friend of both Carrier and Lyle. Like Lyle, Lewis was a Kentuckian and a graduate of the University of Kentucky. He had worked with Willis Carrier in the engineering department of Buffalo Forge Company in 1906–7.

With Carrier, Lyle, Carpender, and Thornton Lewis in agreement, the merger plans proceeded rapidly. By August of 1930 the three companies were exchanging financial statements and by November their stockholders had voted approval. The organization meeting of Carrier Corporation was held on December 11, 1930. Willis Carrier was named chairman of the board; Irvine Lyle, president; Thornton Lewis, executive vice-president. The subsidiary companies included Carrier Engineering Corporation with E. T. Murphy, president, E. T. Lyle and A. E. Stacey, vice-presidents, and L. L. Lewis, secretary; and Carrier-Brunswick International, Inc., with Sidney B. Carpender, president. Charles J. Staples, who had been legal advisor of Carrier Engineering Corporation since its inception, was named secretary and general counsel of the new company.

Carrier Corporation had to survive not only a merger but also the Great Depression. Even before the merger was consummated, Carrier Engineering Corporation had already begun to feel the pinch of hard times. Dividends, which in 1928 had been $2 on the preferred and $8 on the common, were lowered in 1929 and 1930 to $1 on each class of stock. During its first six months Carrier Corporation operated at a loss. Carrier later said:

> That was expected with shift in personnel and realignment of responsibilities that come with a merger. We made a small profit the next six months. From then on, for five years, we lost money—most companies did. But we never lost faith in our ability nor in our product.
>
> Determined to be ready to serve industry and commercial establishments when good times returned, we maintained an active research program through the depression years. We cut all other expenses to the core. Management salaries were reduced. At one time every employee of our company worked without pay for one month to keep the company going. We brought in an especially trained executive, as we thought a businessman could do better than we, who were engineers, in such a time of crisis. But at all times we maintained an active development program. Research is fundamental to our business.

Even before the merger was completed, engineers of Carrier Engineering and of York Heating & Ventilating began co-operative research in Newark. They first concentrated on the development of a room cooler. The result was a unit known as the Atmospheric Cabinet, consisting of a fan, cooling coil, and filter, enclosed in a cabinet, and connected with a refrigerating machine located outside the room to be cooled. Six of the units were sold in early 1931 to the Frigidaire Corporation (now Frigidaire Division of General Motors Corporation) for installation in the offices of Lehman Brothers in New York. The equipment was started up on May 25, 1931, after Frigidaire had changed the refrigerant from sulphur dioxide to "Freon-12," which was destined to revolutionize refrigerating apparatus.

The Atmospheric Cabinet was placed on the market early in 1932. Many were installed to cool offices and residential rooms, but as the depression deepened purchases declined. Nevertheless Carrier Corporation continued to build them and modified their design for one type of today's Weathermaster.

When the demand for individual room air conditioning grew, it was filled with a "self-contained unit," in which the refrigerating machine and air-handling equipment was all housed in one casing. Carrier Corporation brought out its first self-contained unit in 1932, sold a few, then discontinued manufacturing them, as their sale, like that of units with remote refrigerating machines, became the victim of the lengthening depression. There were no important sales of these compact factory-assembled machines until after the middle 1930's. And it was after World War II that there came real volume production of such units as window-sill and floor-model room air conditioners and self-contained "Weathermakers" of larger capacities for stores, shops, restaurants, and suites of offices.

Meanwhile, Willis Carrier turned his attention to "Freon-12." At a meeting of the American Chemical Society in Atlanta in April of 1930, Dr. Thomas Midgley, Jr., had read an historic paper, titled "Organic Fluorides as Refrigerants," which told of the development of "Freon" as a refrigerant for reciprocating machines and described tests conducted at the Frigidaire laboratories in Dayton, Ohio. Carrier and Lyle thought that this refrigerant might be the solution to their problem of getting a safe, small-capacity refrigerating unit. Together with Thornton Lewis, they visited Dr. Midgley's laboratory for first-hand information. Carrier later said:

> I went for one answer and found another of much greater value—an excellent illustration of "serendipity." While there, a chemist described the production of "Freon-12," mentioned the characteristics of a gas obtained in an intermediate step, and stated that there was no intention of producing it except in the industrial process. From the figures he gave me, I believed the gas would be an ideal refrigerant for centrifugal compression, so asked him for the data. They were still written in pencil on work sheets. The chemist had a photostat of them made for me, and later supplied me with a small sample of the fluid. That became Carrene 2, also known as "Freon-11."

The substance obtained in making "Freon-12" was trichloro-monofluoromethane (CCl_3F). Carrier and Stacey started laboratory tests to determine its suitability as a refrigerant for large-capacity centrifugal compressors. Their studies convinced them that the substance had none of the disadvantages of either dielene or Carrene 1 and offered possibilities of greater compression. They believed they could compress the gas from Carrene 2 in two or three stages, instead of four, five, or six when using Carrene 1, and thereby reduce the cost of the compressors by one-half. Carrier made the decision to try it, redesigned the compressor, and assigned the new machine the letter "Z" from "Zeron," the name given the new refrigerant before it finally was called Carrene 2.

86

The new machine was to be manufactured in America but, in June of 1933, Carrier sailed for Germany to get the Jaeger firm to make the pattern drawings. The new machine included a bellows seal which Carrier, characteristically, designed on the boat going to Germany. He did not wait for completion of the new machine to try out the new refrigerant. In July of 1933, Carrene 2 was substituted for Carrene 1 in a 290-ton machine at the Carney's Point plant of E. I. du Pont de Nemours and Company. Greater capacity was produced, operating temperatures reached easier, and the condenser water, which had been inadequate for Carrene 1, proved plentiful for Carrene 2.

The first type-Z machines designed for Carrene 2 were completed in September of 1934 and shipped to the United States Court House in New York. There were three of these, totaling 495 tons capacity for chilling 1,490 gallons of water per minute to 41 degrees F for the building's air conditioning system. In 1950, when the service department of Carrier Corporation was asked for a report on these three machines, no one in the department knew there was such an installation. A service man visited the Court House to check on them and found that the three machines had operated for sixteen years without a service call.

About 270 type-Z machines were sold altogether. In 1937 the company began work on the type-17 machine, with extended surface tubes in the condenser and evaporator. Carrier had suggested such an improvement, using an Aerofin coil, in 1933, but it had not been undertaken because of financial stringencies. The first type-17 was shown at the New York World's Fair in 1939, where it chilled the water to air condition both the Carrier Igloo and the adjacent du Pont building. When the Fair closed, the Melrose Theater in Nashville, Tennessee, purchased this machine which bore the serial number 752. Two years later Carrier's one-thousandth centrifugal refrigerating machine was produced and, on August 4, 1950, the two-thousandth came off the production line.

XII

THANKS to the resourceful mind of Willis Carrier, the steady hand of Irvine Lyle, and the important contributions of scores of other men both in and out of the company, the air conditioning industry by the early 1930's was able to serve all types of buildings except one. The one exception was the most typically American of them all—the skyscraper. Conventional methods in multi-room buildings called for the distribution of conditioned air from central station equipment through large ductwork along walls and ceilings. But the very reason for the skyscraper was to provide more floor space on a comparatively small plot of expensive ground. Hence, ceiling heights and floor areas were of much concern to owners, architects, and consulting engineers. They were chary of sacrificing cube-footage to bulky ducts.

Willis Carrier solved this problem by inventing in 1939 the Conduit Weathermaster System. His patent application for the system which made possible the efficient air conditioning of skyscrapers, new and old, was filed on August 12, 1939, and four separate patents thereon were issued on July 11, August 15, and November 28, 1944. But Carrier's specific interest in this problem reached back to around 1927. Later he recalled how he first turned to skyscrapers:

> During the booming '20s, well before the stock-market crash, one of our sales engineers interested the owner in air conditioning his very tall building on a very small lot in

New York City. The owner's income depended upon rentable space. Determined to use the minimum floor area for risers in the air-distributing system, I considered very high air velocities to keep the duct sizes small. Then came the question of utilizing the energy of the extra-high-velocity air when it reached the room outlets. Its whistling noise would be terrific. Of course, the answer was to apply the ejector nozzle principle we had used successfully for years in dryers. Use less air, discharge it at high velocity through nozzles, set up induced air flow. Nothing came of the inquiry—maybe the owner was hit in the crash.

The inquiry convinced me that soon there would be a market for air conditioning tall buildings and that I had better design such a system. Edible fish were in sight, so I started fishing.

In 1928, while air conditioning the company's own offices in Newark, Carrier designed a system which served as an experimental one for other office installations. The plant was a four-story building. The top three stories were offices so that risers did not constitute a problem as in high buildings. However, eventual solution of the skyscraper problem was under way when this home office installation included window-sill outlets with a heating coil in the connection and an experimental setup for high air velocity in the supply duct to one room. In this duct Carrier enclosed ejector nozzles for induced air flow to air condition the company's board room.

When Carrier put an ejector nozzle inside a duct, he infringed on a patent issued to Dr. Albert Klein on an invention relating to the heating of factory buildings. Klein had put an ejector nozzle inside the duct of a fan-coil heating system to induce a secondary air flow. Carrier later explained:

> Dr. Klein, who at the time of our studies was our European representative, had sold his patent rights to Buffalo

Forge Company. Buffalo had installed only one heating system with the Klein nozzle. We got permission to apply the patent in our air conditioning system, and we called our first apparatus a "high Klein unit." It was actually not a unit as defined today, but a special outlet for central station air conditioning systems. When, later, we applied the ejector nozzle to window-sill-height air discharge, we called it a "low Klein unit."

While the tests in the board room continued, the company sold three systems using the "high Klein unit" to supply the primary air from a central station. These systems were installed in offices in New York, Virginia, and Ohio. The tests included a factor—noise level—which had not been important in industrial installations, where the machines drowned out the air noises, or in drying rooms, which were unoccupied. Air noises were not greatly important even in theater installations until sound was added, thereby compelling the engineers to add sound absorbers in the distribution ducts.

The second major step in the long road toward skyscraper air conditioning came when the company produced a model combining the features of the "high Klein unit" with the under-the-window outlets to produce the "low Klein unit." Carrier later stated:

> With the addition of the ejector nozzles, we obtained induced flow and thereby were able to reduce the quantity of air delivery from the central station apparatus. However, at the time we continued to supply the air through conventional ducts at conventional velocities. The unit became the first model of today's standard Weathermasters, and was similar to the ones used in the Pentagon Building in Washington. Each unit contained a supply connection, return grille to permit induced air flow from the room, nozzles that set up the ejector effect, heating coils in the secondary

air stream to warm the return air, a mixing chamber for the supply and induced air flow, and a discharge grille at window-sill height. None contained any moving parts. All the energy for air circulation was supplied from the fan in the central station system which controlled the primary air—which induced a flow twice that obtained with the Klein unit for heating systems.

The first sale of window-sill-height Weathermaster units was made in March of 1930 to the Superheater Company of East Chicago, Indiana, to condition the air for the second floor of a two-story building. A month later the Phoenix Title & Trust Company of Phoenix, Arizona, bought Weathermasters to air condition its eleven-story building. In August the California Bank of Los Angeles contracted to use 490 Weathermasters in air conditioning its fifteen-story building. More contracts and more research followed—for instance, 79 Weathermasters were sold in 1931 to condition the air in Louisiana State University's Fine Arts Building at Baton Rouge, and 180 for the United States Supreme Court building in Washington. By the end of 1934 twenty-three buildings were using Weathermasters.

Carrier Corporation applied its findings on the development of the Weathermaster to the redesign of the room air conditioners. Thus it was able to reduce their size, quiet them considerably, and raise the ratio of primary to secondary air. By the end of 1936 the company had two methods of air conditioning multi-room buildings. One was the room air conditioner, which cooled, dehumidified, and circulated air drawn from the room or from outdoors. The other was its standard Weathermaster system, which supplied conditioned primary air at conventional velocities from a central station air conditioning apparatus. Carrier later stated:

In contrast to our room cooler, our Weathermasters served the year around; the central station system provided positive control of air in summer and winter. The units contained no moving parts. With them we used smaller central station apparatus than with other types of outlets. With smaller central station air volume, we had reduced the size of the supply risers. But they were still too big for high buildings; they still consumed too much floor space, a disadvantage to all types of applications where maximum rentable or usable area is of prime importance. Often the duct size ruled out the sale of air conditioning systems to owners remodeling multi-story buildings.

The problem, as Willis Carrier saw it in 1937, was to reduce the size of the air risers. To solve it, he returned to the idea that had occurred to him in the booming 1920's—high-velocity air. He began making calculations and designing experimental apparatus. Before he could get his high-velocity-air tests under way, Carrier Corporation decided to move from five plants in four New Jersey and Pennsylvania cities to a single location in Syracuse, New York—occupying long unused factories and offices formerly owned by H. H. Franklin Manufacturing Company. Carrier later wrote:

> When we remodeled the Franklin office building, we installed our experimental apparatus, yet untested, to fill Irvine Lyle's request for the "last word" in multi-room air conditioning apparatus. From that installation came the Conduit Weathermaster System. History repeated itself. The first type of Weathermaster grew from experimental equipment we tried out in our newly acquired office building in 1928; the second from our experimental setup installed in the building we acquired in 1937.
>
> Our 1937 trial system consisted of a central station apparatus which supplied conditioned primary air at high velocity through small rectangular ducts to 43 Weathermasters. They were the cabinet model, and were located under the

windows. In addition to the ejector nozzles to set up the induced air flow, each cabinet housed two coils—one in the primary air stream, the other in the secondary. Each coil performed two functions. In summer, when supplied with refrigerated water, they cooled the air; the same coils, supplied with hot water in winter, heated the air.

By having the same coils perform two functions, the unit was kept small enough to install inside the wall under the window with the discharge outlet in the sill. This feature enables us today to offer two types: the cabinet, which encloses the integral parts and is located in front of the window; and the wall, with the parts built into the building structure. The first was particularly suitable for installation in existing buildings, the wall-type applicable to new construction. . . .

One defect quickly showed up in the test installation—the ducts leaked. The air was distributed at high velocity, which meant high pressure in the ducts, and only about 50 percent of the volume discharged by the fan was reaching the Weathermasters. Carrier saw that much research was needed to improve the ducts. The first change in the ducts was made early in 1938 when Carrier Corporation made the first sale of the new system to the Marlyn Apartments, then under construction in Washington, D.C. Carrier said:

In order to cut down leaks in the air-supply ducts, we made them round. They leaked some, but were tighter than our rectangular ones, even after puttying and tightening the joints. But they were costly.

We made another change. The central station apparatus, which handled all outside air required for ventilation, conditioned the primary air for the Conduit Weathermaster with one exception—dehumidifying. We calculated that the units would remove all of the excess moisture from both the primary and secondary, or recirculated, air. They took out the moisture all right, so much of it that the drip pans spilled over and ruined the parquet flooring. It cost us plenty to repair the floors, but correcting the installation was simple—we chilled the water in the central station spray apparatus, thus dehumidifying the primary air before it entered the Weathermasters. From then on we had no trouble.

To make the ducts less costly, Carrier changed to round steel tubes and standardized their length for minimum field fabrication. He tested various means for joining the lengths, analyzed friction losses, estimated expansion and contraction with temperature changes which could be considerable in a riser for, say, a 20-story building. By the fall of 1938 his plans were sufficiently defined to present the patentable features to the company's attorneys.

While working on redesigning the system Carrier asked for criticisms from the company's sales engineers and construction superintendents. When he had three models ready for demonstration, he invited these field men to visit the laboratory and inspect them. By the end of 1939 Carrier's new design for the high-velocity-air-supply Weathermaster system was complete. It included many of the features developed to bring prices down and still maintain high performance specifications. The first sales of the new system were made in 1940 for installation in the Bankers Life Building, Macon, Georgia, the Durham Life Building, Raleigh, North Carolina, and the United Carbon Building, Charleston, West Virginia. Each fulfilled the goal which Willis Carrier first set in 1928 and on which he began concentrating his efforts in 1937: an air conditioning system suitable for multi-story, multi-room buildings.

More sales of the revolutionary system were made before America went to war in 1941, among the most notable of which was the Statler Hotel in Washington, D. C., the last hotel in the nation to be built before the opening of hostilities. Here the installation of a Weathermaster system using conduit instead of conventional ducts saved the equivalent of two floors of rentable space.

More refinements came after victory in 1945. In 1952, when Carrier Corporation was celebrating the Golden Anniversary of Air Conditioning, no monument to Willis Carrier eclipsed his Conduit Weathermaster System. One of the most talked-of installations was in the 40-story Secretariat building of the United Nations and the world's largest was that for the Gateway Center group of office buildings in Pittsburgh, requiring 4,500 tons of refrigeration and over 6,000 Weathermaster units in individual offices. In these installations, outdoor air is washed, filtered, and humidified or dehumidified by central station equipment to meet the needs of the season. Conditioned air is then distributed at high velocity through conduits to various floors and rooms. Here the conditioned air is discharged and diffused throughout the rooms at the same time that room air is drawn over heating or cooling coils of Weathermaster units beneath the window sills. The advantages of this system include the space-saving elimination of bulky ducts, the provision of heating, cooling, humidity control, and ventilation with a single system, and the use of a simple dial control in each room for the selection of individual room temperatures.

XIII

DURING World War II the nation's production was greatly aided by the industry Willis Carrier fathered, by the company he and Irvine Lyle created, and by Carrier's personal engineering talents.

The need for turning out more goods faster than at any time in America's history was translated not only into a call for air conditioning but also refrigerating equipment capable of producing extremely low temperatures. These were especially needed in the synthetic rubber and high octane gasoline programs so vital to war machines. At one point, it was necessary to remove Carrier heavy-duty centrifugals from great stores such as Tiffany's, Hudson's, Lord & Taylor's, and Macy's for installation in war production plants. Carrier air conditioning and refrigerating equipment was required for warships and cargo vessels, for munitions plants, and for factories specializing in the production of such essential war material as bombsights and other precision instruments. Thousands of walk-in coolers for food storage were ordered by the armed forces for use both at home and in war zones. Special portable coolers were manufactured to permit the servicing of airplanes in hot climates. Air conditioning units were produced for military-photographic and bombsight-repair trailers.

The war work of Carrier Corporation was not confined to products in its own field. The company turned out airplane engine mounts, sight hoods for guns, tank adapters, and other military and naval items. However, its principal assignment outside the air conditioning field involved the redesign and exclusive production of the "Hedgehog," a device for discharging 24 antisubmarine bombs simultaneously in a prearranged pattern.

In recognition of its over-all contribution to the war effort, Carrier Corporation was awarded the Army-Navy "E" six times, an honor attained by only thirteen other companies.

Irvine Lyle died early in the war, on June 7, 1942. His age was sixty-eight years, of which forty had been devoted to air conditioning. He was succeeded as president by Cloud Wampler, forty-seven-year-old Midwestern businessman who had been a director of the company and executive committee member since 1935, chairman of the finance committee since 1938, and executive vice-president since September 1, 1941.

At the time of Lyle's death, Willis Carrier, then sixty-six and past the so-called "retirement age," was engaged in what he later referred to as his greatest engineering achievement. Shortly before his own death Carrier stated:

> Once, I accomplished the impossible. That is, the task seemed impossible when I first tackled it. And because of its success, high officials in the Air Force told me that World War II was shortened by many months.

Others familiar with his work agreed that there never was a more difficult, more exacting, or more vital air conditioning and refrigerating system than the one designed for the National Advisory Committee for Aeronautics and installed in its wind tunnel at Cleveland, Ohio, to simulate freezing high-altitude conditions for the testing of prototype planes.

When the N.A.C.A. proposed this wind tunnel in 1940, nothing comparable in size had ever been considered. In it the complete engine assembly and propellers would be tested under flying conditions. Ten million cubic feet of air per minute had

97

to be cooled to a temperature of 67 degrees F below zero. Carrier believed that his centrifugal refrigerating machines would be particularly advantageous. He also believed he could be of some use in selecting the cooling equipment, as his calculations indicated that standard coils required to cool the air would not fit into the space. After Congress approved the expenditure of $4,900,000 on the wind tunnel in July of 1941, government engineers drew up plans and specifications for an experimental cooling coil using streamline tubes.

Carrier believed they were on the wrong track. So he and his engineers began building their own test apparatus to secure data that would prove the superiority of the coil he recommended and provide convincing arguments for the abandonment of the streamline tubes. Meanwhile N.A.C.A. was conducting experiments on the streamline tubes at Langley Field, Virginia. Carrier went to Washington, called on Dr. Vannevar Bush, Director of the Office of Scientific Research and Development, and had him arrange a lunch with Dr. Jerome Hunsaker, chairman of N.A.C.A., who brought with him Dr. George W. Lewis, N.A.C.A.'s director of research. Carrier later told of the luncheon:

> Dr. Lewis asked me if I thought the tests on the streamline coils at Langley Field had value. My answer was not polite, and I'm afraid I scared our representative by my outburst. I told Dr. Lewis that the boys conducting the tests did not know what it was all about, and that too much money and, of more importance, too much time had been wasted already. "Heat transfer experts should be called in," I told him and suggested, among others, Professor William H. McAdams of Massachusetts Institute of Technology.

Carrier came home convinced his recommendations for the cooling coil would be considered, and therefore concentrated on its design for the wind tunnel.

The research involved two inter-related problems—the cooling coil and the refrigerating system. The amount of heat removed from the air blown over the coils depended upon the capacity of the refrigerant inside the tubes to absorb the heat. As "Freon-12" which was specified as the refrigerant had never been used in any sizable system to reach the low temperature of minus 67 degrees F, the coil tests involved basic research on the refrigerant itself. This in turn involved the design of the refrigerating system. Of the two inter-related problems, that posed by the cooling coils was the more difficult. Carrier later said:

> Calculations indicated we would need a direct expansion coil with a face area of approximately 8,000 square feet. The wind tunnel, 51 feet in diameter, had only 2,000 square feet of cross-sectional area. Quite a feat to fit 8,000 square feet into 2,000! Yet the solution was actually simple to accomplish. We jackknifed the sections, folding them down like a collapsed accordion until the coils fitted into the tunnel.

There were many questions on which no data were available. To answer them a miniature wind tunnel was built on the mezzanine floor of Carrier Corporation's power plant. As a result of tests in this tunnel, Carrier and his associates found a way to secure vaporization of the refrigerant throughout the full length of the cooling coil. They did it by distributing the refrigerant in such quantities and at such pressures that there would be an excess of liquid for each tube.

By January of 1942, Carrier engineers had redesigned their centrifugal compressor for "Freon-12." The fourteen 1,500-horsepower refrigerating machines, in addition to maintaining conditions of air simulating altitudes up to 30,000 feet, had to cool fifty pounds of gasoline per minute for the engines, cool the make-up air to the tunnel, chill water, and refrigerate the coils for an icing tunnel located nearby when the wind tunnel was shut down.

Bids were opened on March 4, 1942, and on March 16 Carrier

Corporation was awarded contracts for both the cooling coils and the refrigerating system. Then began the exacting work of testing many component parts. Carrier said:

> Much was not standard, nor could it be, for such an unusual installation. We planned on using many new devices, any one of which could cause failure of the entire system. To guard against such catastrophes, we carefully checked everything before shipment. For example, we tested approximately 12,000 tubes, each fitted with a turbulator. ·
>
> In developing new equipment for the wind tunnel, from weighted valves in the refrigerant circuit to suction dampers for controlling capacity, we followed a principle which I found from experience was a wise one. We researched and built the valves ourselves. We had to have them to make the job work; later, should a market develop for them, we would turn our drawings and specifications over to a manufacturer of valves.

More than a year passed before the entire tunnel was ready for operation. After many shake-down difficulties and numerous false starts, the system was ready on April 24, 1944, for a formal run-in test. Carrier was there for the start up and every one of his engineers who had worked on the job was assigned a "battle station." In short order N.A.C.A. knew it had what it needed to help win the war.

After the war Willis Carrier announced that he was going to rest. So he set out with Mrs. Carrier—his third wife, whom he had married in 1941—on a three-month trip to South America. Everywhere they went in their tour, which covered thirty cities, people flocked to pay him homage. Leading industrialists, scientists, and educators honored him at banquets and receptions. It was proof of the power of the idea he had fathered which in forty-five years had crossed all international boundaries.

For the next three years Willis Carrier followed a routine that revealed the inner courage of the man. At doctors' orders he stayed horizontal twenty hours a day because of a heart ailment. But his enforced physical inactivity never quieted his restless mind. In February, 1948, he was made Chairman Emeritus of Carrier Corporation. In 1949 he was still coming regularly, though less often, to the office and his home visiting list was long. Carrier engineers were frequently in touch with him. Mainly, though, he was on his back, a pad of paper on his knees, his slide rule close at hand, figuring out ways to simplify complex calculations or to reduce vague concepts to concrete terms.

In September, 1950, he said:

> My routine is broken at intervals with trips to New York Hospital for check ups. I am due to go there later this month, but have advanced the time to coincide with a lecture at Columbia University. Dr. Richard Planck, an internationally famous refrigerating engineer, is to be the speaker.

It was his last journey in quest of knowledge. Willis Haviland Carrier died in New York on October 7, 1950, shortly before his seventy-fourth birthday. He had pioneered an industry, helped create and build up an enterprise, and measurably advanced scientific knowledge. "Father of Air Conditioning" is a title that fitted him well!

XIV

THE story of any man, in its final writing, must give attention to the length of the shadow which he cast: to enduring contributions which do not cease with his passing.

It must look beyond his personality, the day-to-day happenings of his life, and regard his ideas and inventions not in the light of immediate accomplishments so much as in terms of long-range results.

So regarded, the monument to Willis Carrier is the industry he founded and, more especially, the company which bears his name.

Half a century has passed since the exploring mind of Willis Carrier conceived the idea that enabled man to do something about the weather. Yet even in this space of time, short as years go, the Carrier concept of controlling indoor climate has brought about many changes in our way of life.

Free of depressing heat and sticky humidity, a man can now sleep soundly through sultry summer nights and work at top efficiency through hot days. His wife favors shops which consider cool comfort as necessary as show windows. He lunches in restaurants where heat cannot dull appetite. The hotels where he stays offer more than a bed for the night. His hospitals contain a better atmosphere for convalescence. His entire life is a better life.

The same air conditioning which enables man to live more comfortably enables him to enjoy better products, too. From the nylon bristles in the toothbrush he uses in the morning to the alarm clock he sets at night, thousands and thousands of everyday things have been produced better or cheaper because of air conditioning. Without air conditioning, indeed, many of them could not have been produced at all.

As might be expected, both by reason of its seniority and its constant emphasis on research and progressive engineering, the Corporation which bears the Carrier name has made many of the most important air conditioning installations—as well as the first—in almost every field. In our country, equipment designed and built by Carrier Corporation brings comfort to those who live, work, or relax in such famous buildings as the United States Capitol, the Pentagon, Radio City Music Hall, the new Statler Center in Los Angeles, Macy's, the Mayo Clinic, the United Nations Secretariat, and Lever House. The latter two are among nine great Carrier air conditioned office buildings which have been erected in New York City since the close of World War II.

World travelers will find Carrier systems in such well-known buildings as the Imperial Hotel in Tokyo, the Bank of India in Bombay, Broadcast House in London, the Municipal Auditorium in Mexico City, the Caribe Hilton Hotel in San Juan, and MacDonald House in Singapore. Or in the Houses of Parliament of Canada, Egypt, India, Lebanon, Norway, Peru, and South Africa.

On the high seas, six of the ten largest passenger liners afloat—including the *Queen Elizabeth*—are Carrier-equipped, as will be the world's greatest aircraft carrier, the *U.S.S. Forrestal*.

In recent years, so-called *big* air conditioning has come even more into its own. Relatively few large buildings are designed without air conditioning. But it is in the area of *small* air conditioning that the growth curve has risen most sharply. For today air conditioning is available in packages. There are small units for individual rooms. There are somewhat larger units for shops. There are other compact units that provide year-round air conditioning for the entire home. Everything in the way of air conditioning equipment, from the smallest to the largest, is now available—a room air conditioner with a capacity of only one-third ton and then step-by-step up to a central installation measured in thousands of tons; tiny air conditioning packages at one end, giant systems at the other, with each size space in between adequately provided for.

This wide acceptance of air conditioning to provide human comfort has often overshadowed the vital importance of its original and basic contribution: service to production. In more than two hundred industries, Carrier research and technical ability is demonstrating that vagaries of weather or atmosphere can be controlled to meet the most severe specifications of modern manufacturing.

Accurate control of humidity is required in industries which process hygroscopic materials such as candy, food products, textiles, paper, and pharmaceuticals. Here air conditioning is able to maintain uniform weight, dimension, and strength, improve workability and the ease of handling, and insure high speed and constant production rates.

Air conditioning has also become increasingly important to industries which utilize non-hygroscopic materials—those in which the regulation of temperature and humidity is required to control surface film, reduce abrasive dusts and corrosion, and maintain close tolerances. Many of these applications are in laboratories and machine shops but they are also in factories turning out cameras, electronic devices, optical goods, and plastics.

Air conditioning is likewise finding wider and wider usefulness in industries where the control of chemical reactions is required. Among numerous installations of this nature are those in blast furnaces, dye works, and laboratories—and in plants producing explosives, heavy chemicals, rayon, and rubber.

A fourth type of industry which uses air conditioning—and one which has made tremendous strides within recent years, particularly in respect to medicine—is that involving biological or bacteriological processes. In this field air conditioning is able to control the rate of growth and the character of bacteria, molds, and enzymes, as well as deter decay and spoilage, and prevent contamination. Such applications include antibiotics and biologicals, but they are also found in breweries, distilleries, and in the food and meat-packing industries.

Among the hundreds of Carrier customers are twenty of the industrial giants of America. Ten of these concerns have been served by Carrier for twenty-five years or longer, five for twenty or more years, and the remaining five for better than ten years. Such a record is tribute indeed to Willis Carrier and his fellow engineers.

And yet gaining such acceptance for air conditioning was a tedious process. For example, the growth of Carrier Corporation, which for years represented the major part of the fledgling air conditioning industry, was slow, unspectacular, but consistent prior to World War II. "Air conditioning is right around the corner," most people said, feeling that an industry with such potentials would soon capture the public imagination.

The turn came during World War II when the "luxury label" that had been attached to air conditioning gave way to an appreciation of its practical use in factories as an aid in making goods better, faster, and cheaper—a concept which also included increased production resulting from improved human efficiency.

After the cessation of hostilities in 1945, air conditioning broke out of the development era in which all new industries traditionally linger and its growth curve has pushed rapidly upward ever since.

Some indication of the recent growth of the air conditioning industry may be found in the Carrier Corporation record. Completed sales increased from approximately twenty million dollars in 1942 to fifty-three million dollars in 1947, and then to more than one hundred million dollars in 1952—over a span of just ten years. Production facilities were less than two million dollars in 1942 and over eighteen million dollars a decade later. Even more significant, Carrier now provides over 7,000 men and women with the means of making a living. And many of these have spent their entire working lifetimes mastering the art and the science of air conditioning.

Much of this progress revolves around three men—Willis Carrier, the inventor and engineering genius; Irvine Lyle, the man with the business vision; and Cloud Wampler who took the helm in 1942. Each of these made his special contribution and the merging of their efforts through the capacity and devotion of many others is responsible for the Carrier Corporation of today.

It is the air conditioning industry—with the Carrier enterprise as pioneer—that is the real monument to Willis Carrier. But this monument is not a static thing. For both the industry and the enterprise are dynamic. And there will be no limit to the height and width and breadth of the monument.

Willis Carrier had many dreams for the industry he founded. Some of these seemed almost fantastic at the time. Most of his dreams came true, however—and during his lifetime. A few—like the air conditioned streets he once prophesied and the air conditioning of whole cities from a central plant—have not come yet.

But who can say that, in the air conditioned world of tomorrow, they will not come? And even greater things than these!

CHRONOLOGICAL TABLE OF EVENTS WHICH LED TO MODERN AIR CONDITIONING

1500–1952

"Air conditioning is the control of the humidity of air by either increasing or decreasing its moisture content. Added to the control of humidity is the control of temperature by either heating or cooling the air, the purification of the air by washing or filtering the air, and the control of air motion and ventilation."

WILLIS H. CARRIER
FEBRUARY 28, 1949.

RELATED DEVELOPMENTS AND THEORIES	HEATING AND VENTILATING FANS, HEATERS, HEAT PUMPS	REFRIGERATION (NOT INCLUDING COMMERCIAL EQUIPMENT AS DOMESTIC REFRIGERATORS, ETC.)	AIR CLEANING, HUMIDIFYING, PURIFYING. WASHERS, HUMIDIFYING UNITS OR HEADS, ODOR ABSORBERS, ETC.	AIR COOLING. FAN-ICE, COOLING-COIL AND FAN-COOLING-COIL UNITS AND SYSTEMS	AIR CONDITIONING. SPRAY TYPE CENTRAL STATION APPARATUS AND CHEMICAL DEHUMIDIFIERS
YEAR	YEAR	YEAR	YEAR	YEAR	YEAR
1500 (Circa) Leonardo de Vinci (1452–1519) Devised instrument for measuring changes in moisture in air. (1) Devised instrument for measuring wind velocities. (1)	1500 (Circa) Leonardo de Vinci. Built ventilating fan, water driven to ventilate boudoir of wife of his patron. (1)				
1607 Galileo Galilei (1564–1642) Invented a thermometer. (2)					
1643 Torricelli (1608–1647) Invented a barometer. (2)					
1659 Robert Boyle (1627–1691) British Scientist. Discovered relationship between density of air and its temperature. (2)	1660 Sir Christopher Wren (1632–1723) English Architect. Designed gravity exhaust ventilating system for House of Parliament. (3)				

1714 Gabriel Daniel Fahrenheit (1686-1736) German Physicist. Introduced Fahrenheit thermometer. (2)

1730 (Circa) Henri Y. Pitot (1695-1771) Invented Pitot tube. (4)

1742 Andres Celsius (1701-1744) Introduced Centigrade thermometer. (2)

1783 Antoine Laurant Lavoisier (1743-1794) Established true nature of atmospheric air. (2)

1797 Benjamin Franklin asked, "Whence comes the dew that stands on the outside of a tankard that has cold water in it in summertime?" from: "Benjamin Franklin" by Carl Van Doren, Viking Press, New York, 1938, pg. 77

1800 John Dalton (1766-1844) English Chemist and Physicist. Formulated laws of pressure of water vapor in air. (2)

1820 Wollmann or Kallstenius probably employed an anemometer for measuring air flow. (10)

1736 Dr. J. T. Desagulier (1683-1744) French Naturalist. Designed centrifugal blowing wheel, manually operated. Connected it to gravity exhaust ventilating system in House of Commons. (5) For manually operated fan, "An ordinary able-bodied man can easily exert 3000 ft. lb. per min." (6)

1812 Daniel Pettibone, Philadelphia. Invented means for "Warming Rooms by Rarefied Air." No patent number. May 12, 1812.

1814 Marquis of Chambannes. Wrote "On Conducting Air By Forced Ventilation." Stated he hoped to be known as author of forced ventilation. (5)

1775 Dr. William Cullen (1710-1790) Scottish Physician, University of Edinburgh, Wrote "Essay on Cold Produced By Evaporating Fluids." (8) Reduced atmospheric pressure in a vessel of water with air pump to produce vacuum and freeze the water. (9) 1785 J. & E. Hall, Ltd., Dartford, Kent, England. Established in 1785. Began selling refrigerating machines in 1881.

1823 Sir Humphrey Davy (1778-1829) and Michael Faraday (1791-1867) Demonstrated that gases could be liquefied by pressure. (9)

1771 G. Bacarria Recorded observations on electric discharge through smoke filled gases. (7)

1819 Rafinesque Wrote article on electrical precipitation in American Journal of Science and Arts. Theory applied in 1880 by K. Moeller of Brachwede, Germany, but not successfully. (7)

RELATED DEVELOPMENTS AND THEORIES	HEATING AND VENTILATING FANS, HEATERS, HEAT PUMPS	REFRIGERATION (NOT INCLUDING COMMERCIAL EQUIPMENT AS DOMESTIC REFRIGERATORS, ETC.)	AIR CLEANING, HUMIDIFYING, PURIFYING. WASHERS, HUMIDIFYING UNITS OR HEADS, ODOR ABSORBERS, ETC.	AIR COOLING. FAN-ICE, COOLING-COIL AND FAN-COOLING-COIL UNITS AND SYSTEMS	AIR CONDITIONING. SPRAY TYPE CENTRAL STATION APPARATUS AND CHEMICAL DEHUMIDIFIERS
YEAR	YEAR	YEAR	YEAR	YEAR	YEAR
	1824 Nicholas Leonhard, Sadi Carnot (1796–1832) French Physicist. Propounded theory of heat pump. (11)	1824 Heat pumps, (see Column 2) 1824. 1824 John Vallance, Brighton, Sussex, England. Patented sulphuric acid absorption process of refrigeration. (8)	1824 Hohfield Suggested use of electricity to precipitate dust in atmosphere. (7)		
	1832 George Harley and John Sedgwick, Philadelphia. Invented "air or ventilator pump." Listed in U. S. inventions 1790–1847. No patent number—papers destroyed in patent office fire 1835.				
1836 James Apjohn (1796–1886) Irish Chemist. University of Dublin. Propounded theory of adiabatic absorption of moisture by air but was unable to establish its correctness. (13)		1834 Jacob Perkins (1766–1849) Newburyport, Mass. First in America to patent a refrigerating machine (sulphuric ether, closed cycle, compression.) Machine not commercial success. (8) U. S. Patent #6662 issued August, 1834.		1833 Dr. John Gorrie (1803–1855) American Physician, Charleston, S. C. Hung buckets of ice in hospital rooms, blew air over ice to cool rooms for malaria and yellow fever patients. (Statue in U. S. Hall of Fame, Washington, D. C.) (12)	
1843 David B. Van Tuyl, Dayton, Ohio. Invented "Instrument for Regulating Temperature." No patent number. October 12, 1843.	1839 M. Combe. Designed multi-blade centrifugal fan with curved blades for ventilating Belgian mines. (15)		1838 Dr. David Boswell Reid (1805–1863) Added moisture to air with perforated pipes in forced ventilating system of House of Commons. He recommended 30 cfm of air per person, also recommended a cloth filter to clean air drawn from outdoors. (14)	1845 Prof. R. Ogden Doremus (1824–1906) Professor of Chemistry, College of City of New York. Wrote article on cooling hospitals. (16)	
1847 James Glaisher (1809–1908) Engineer, Astronomer, Meteorologist. Computed reliable stationary hygrometer tables. (Made balloon ascents 1862–1866 to obtain meteorological data.) (17)	1847 W. Buckel. Presented design data for fans before Institute of Mechanical Engineers. (18) Data continued in use by many engineers as late as 1900. (19)	1849 Dr. John Gorrie. (See Column 5, 1833). Invented ice machine. Forerunner of today's compressed air refrigerating machines. Issued May 6, 1851. U. S. Patent #8080. (9) (12)		1848 Dr. David Boswell Reid. (See column 4, 1838) Suggested circulating artesian well water through steam heating pipes to cool House of Commons in summer. (14)	

1850 Guiyard
Observed wire charged with static electricity plunged into smoke filled jar, cleared the air in the jar. (7)

1850 Several countries established a network of weather stations. Meteorological offices in London 1854. (2) In United States 1870. (20)

1851 Taylor Instrument Companies, Rochester, New York. Established 1851.

1851 Ferdinand P. E. Carré, France. Designed first ammonia absorption refrigerating machine. (2) U.S. Patent #30201. Issued October 2, 1860. (8)
1853 Alexander C. Twining (1801–?) New Haven, Connecticut. Invented sulphuric ether compression refrigerating machine. U.S. Patent #10221. Issued November 8, 1853. Was reported in 1856 as making ice at rate of 2000 lbs. in 24 hrs. (8)
1853 Frick Company, Waynesboro, Pennsylvania. Established in 1853.

1854 (Circa) Lord Kelvin (William Thomson) (1824–1907) British Physicist. Further advanced theory of heat pump advocated by Carnot 1824. (11)

1854 Heat Pump (See preceding column.)

1855 J. R. Barry. Invented railroad car ventilator. Fan blew air across tank of water for cleaning. U.S. Patent #12851. Issued May 15, 1855.

RELATED DEVELOPMENTS AND THEORIES	HEATING AND VENTILATING FANS, HEATERS, HEAT PUMPS	REFRIGERATION (NOT INCLUDING COMMERCIAL EQUIPMENT AS DOMESTIC REFRIGERATORS, ETC.)	AIR CLEANING, HUMIDIFYING, PURIFYING, WASHERS, HUMIDIFYING UNITS OR HEADS, ODOR ABSORBERS, ETC.	AIR COOLING. FAN-ICE, COOLING-COIL AND FAN-COOLING-COIL UNITS AND SYSTEMS	AIR CONDITIONING. SPRAY TYPE CENTRAL STATION APPARATUS AND CHEMICAL DEHUMIDIFIERS
YEAR	YEAR	YEAR	YEAR	YEAR	YEAR
	1857 Prof. (William John Macquorn) Rankine (1820–1872) Scottish Engineer and Physicist. One of the founders of modern science of thermodynamics. Designed fan with spiral or scroll shape housing. (18)	1856 Au Sieur Baudelot à Harancourt. Invented Baudelot coils for a refrigerating system. Coils were later employed to chill water for air-conditioning systems. (See Carrier Patents, U.S. #1,078,-608, January 19, 1912.) Baudelot coil, French Patent #16,167 was designed to chill all liquids, especially beer. (21)		1856 Azel S. Lyman, New York. Invented "Method of Cooling and Ventilating Rooms." Air blown over ice in racks at ceiling of room. U.S. Patent #14,510 issued March 25, 1856. Extended, then reissued 1874. (See below, 1874.)	
	1860 B. F. Sturtevant Company (now Sturtevant Division, Westinghouse Electric Corporation), Boston, Massachusetts. Began building centrifugal fans. Issued catalogue in 1866. Published rating tables for various operating conditions of fans in 1876 1862 Hon. Henry Ruttan, Ruttan Warming and Ventilating Company, Chicago. Wrote "Warming and Ventilation of Buildings," copyrighted 1888.		1860 T. E. McNeil. Invented air moistening or humidifying arrangement. Pans of water in hot air registers. U.S. Patent #27461. Issued March 13, 1860. 1861 G. E. J. Colburn. Invented air moistener for floor type hot air registers. U.S. Patent #31,152. Issued January 22, 1861. 1861 F.H. Furness. Invented railroad car ventilator with scrubbing vanes. U.S. Patent #33,302. Issued September 17, 1861.		

1865 William Edson, Boston, Massachusetts. Invented a "Hygrodeik," an instrument for measuring moistness and dryness of air. Consisted of wet- and dry-bulb thermometers and chart for reading "the state of the air we breathe." Instrument mentioned in Oliver Wendell Holmes' "Guardian Angel" 13th Chapter.

1870 B. F. Sturtevant, Jamaica Plain, Massachusetts. Invented "Improvement in Compound Air-heaters and Steam Condensers." Centrifugal type fan and coil unit heater. U.S. Patent #100,241. Issued February 22, 1870.

1865 Guibal Designed multi-blade centrifugal fan with chimney or discharge tube, a feature used in mine ventilation as late as 1915. (18)

1866 Thaddeus Lobieski Constantine Lowe. (1832-1913) American aeronaut, interested in ballooning. Developed closed cycle carbon dioxide refrigerating machine. "Mode for Manufacturing Ice," U.S. Patent #63,413. Issued April 2, 1867. Installed machine on freighter "Taber" 1867. (8)

1867 The Vilter Manufacturing Company, Milwaukee, Wisconsin. Established in 1867. Began selling refrigerating machines in 1885.

1870 Franz Windhausen, Brunswick, Germany. Designed compressed air refrigerating machine. (8)

1866 J. D. Whipley and J. J. Storer, Boston, Massachusetts. Invented "Apparatus for Removing Dust from Air." U.S. Patent #53,068. Issued March 6, 1866.

1866 Dr. Joseph Lister (1827-1912) English Surgeon. Noted that carbonic acid was used to deodorize sewage in Carlisle. Adapted practice to purify air in operating rooms. Later, 1869, sprayed watery solution of carbonic acid in air of operating rooms. Procedure led to antiseptic practices. (22)

1867 D. E. Somes, Washington, D.C. Invented "Air Moistening, Cooling and Warming" apparatus. U.S. Patent #61,886. Issued February 5, 1867.

1865 N. S. Shaler, Newport, Kentucky. Invented "Air Cooling Apparatus." U.S. Patent #47,991. Issued May 30, 1865.

RELATED DEVELOPMENTS AND THEORIES	HEATING AND VENTILATING FANS, HEATERS, HEAT PUMPS	REFRIGERATION (NOT INCLUDING COMMERCIAL EQUIPMENT AS DOMESTIC REFRIGERATORS, ETC.)	AIR CLEANING, HUMIDIFYING, PURIFYING. WASHERS, HUMIDIFYING UNITS OR HEADS, ODOR ABSORBERS, ETC.	AIR COOLING. FAN-ICE, COOLING-COIL AND FAN-COOLING-COIL UNITS AND SYSTEMS	AIR CONDITIONING. SPRAY TYPE CENTRAL STATION APPARATUS AND CHEMICAL DEHUMIDIFIERS
YEAR	YEAR	YEAR	YEAR	YEAR	YEAR
		1870 U.S. Census reported four plants in United States making ice. Total investment $434,000, payroll $40,000 annually, business $258,230. (8) 1872 David Boyle (1837-?). Johnstown, Scotland and Mobile, Alabama. Invented ammonia compression refrigerating machine. Called "Father of ammonia compression refrigeration." Produced commercial ice October, 1873. (8) U.S. Patent #128,488. Issued June 25, 1872. Prices of complete machines and ice plants of that period: 1 ton capacity = $ 4,000.00 10 ton capacity = 16,000.00 30 ton capacity = 40,000.00 A 30 ton plant was considered large up to 1890. (23)	1871 James G. Weldon, Pittsburgh, Pennsylvania. Invented method to introduce steam in air passages of air furnace to moisten air. U.S. Patent #119,955. Issued October 17, 1871. 1872 Levi K. Fuller, Brattleboro, Vermont. Invented ventilator and dust arrestor. Fan—blades dipping in water tank—blew air across turbulent surface for cleaning action. U.S. Patent #131,266. Issued September 10, 1872. 1872 G. M. Parks, Fitchburg, Massachusetts. Started firm which became G. M. Parks Company in 1901 and in 1918 purchased Stuart W. Cramer interest to form Parks-Cramer Company, Fitchburg, Mass. and Charlotte, N. C.	1871 Andrew Muhl, Waco, Texas. Invented "Apparatus for Cooling the Air in Buildings." Blew air from forced ventilating system through "conduits or tubes" over refrigerated coils placed near ceiling in each room. U.S. Patent #146,267. Issued January 6, 1874.	

1873 U.S. Patent Office Index from 1790–1873 lists inventions of "Ventilators," a total of seventy-six.

1873 Prof. C. P. G. Linde, Munich, Germany. Introduced ammonia refrigerating machine in 1873–1875. (9) U.S. Patent #228,-364. Issued June 1, 1880.

1873 In U.S. Patent Office Index from 1790–1873 no invention was listed as air moisteners or air washers.

1874 Azel S. Lyman, New York. Invented a machine for purifying, drying, and cooling or warming air. Passed air through beds of wetted charcoal. (16)
Also invented "Method of Cooling and Ventilating Rooms." Air blown over ice in racks at ceiling of room. Original patent issued 1856 (see above). U.S. Patent Reissue #5,786. Issued March 10, 1874.

1874 York Corporation (Originally York Ice Machine Company) York, Pennsylvania. Founded in 1874. Began building refrigerating machines in 1885.

1875 International Congress fixed various standards of moisture in a number of textiles. Established uniform system of numbering yarns. (24)

1876 Thomas L. Rankin. Invented absorption type refrigerating machine used in breweries, packing houses, refrigerated cars, skating rinks. Between 1868 and 1884 twenty-four patents were issued to him. First on ice machine, U.S. Patent #175,498. Issued March 28, 1876. (9)

RELATED DEVELOPMENTS AND THEORIES	HEATING AND VENTILATING FANS, HEATERS, HEAT PUMPS	REFRIGERATION (NOT INCLUDING COMMERCIAL EQUIPMENT AS DOMESTIC REFRIGERATORS, ETC.)	AIR CLEANING, HUMIDIFYING, PURIFYING. WASHERS, HUMIDIFYING UNITS OR HEADS, ODOR ABSORBERS, ETC.	AIR COOLING. FAN-ICE, COOLING-COIL AND FAN-COOLING-COIL UNITS AND SYSTEMS	AIR CONDITIONING. SPRAY TYPE CENTRAL STATION APPARATUS AND CHEMICAL DEHUMIDIFIERS
YEAR	YEAR	YEAR	YEAR	YEAR	YEAR
1878 Upward vs. downward air flow in ventilating halls—arguments summarized in Report No. 119 of Documents of the House of Representatives (U.S.) in 1879. (3)	1878 Buffalo Forge Company, Buffalo, New York. Established in 1878. 1878 L. J. Wing, founder of L. J. Wing Mfg. Co., Linden, N.J. was awarded several medals for "Promotion of Mechanics" with his design of the Wing Disc Fan. U.S. Patent #215,783. Issued May 27, 1879.	1877 Franz Windhausen, Berlin, Germany. Invented water-vacuum refrigerating machine. Patented in Germany, December 14, 1877. U.S. Patent #236,471. Issued January 11, 1881. 1879 John C. De La Vergne and William M. Mixer. Invented an ammonia compression machine with improved sealing, oil pumping, etc. Installed machine in Herman Brewery, New York. (9)	1879 The atomizing nozzles invented by J. G. Garland installed in cotton mill owned by J. F. Slater. Added moisture reduced electricity in cloth which formerly was sufficient to cause "your beard and hair" to stand out straight. (Atomizing nozzle invented by J. G. Garland in 1880, see below.) (2)		

1880 American Society of Mechanical Engineers organized 1880.

1881 American Blower Corporation (originally Huyett & Smith Manufacturing Company), Detroit, Michigan. Established in 1881.

1880 De La Vergne Refrigerating Machine Company, New York. Established 1880. In 1909 became De La Vergne Machine Co. Later The Baldwin Southwark Corp., Philadelphia, a subsidiary of The Baldwin Locomotive Works, Philadelphia.
1880 Henry Vogt Machine Company, Louisville, Kentucky. Established in 1880. Began building refrigerating machines for ice plants in 1885.

1880 Emil H. C. Oehlmann, Berlin, Germany. Invented air moistening head. Later manufactured by American Moistening Company, Providence, Rhode Island, now a division of Grinnell Company. German Patent #12,520. Issued March 26, 1880. U.S. Patent #267,102. Issued November 7, 1882.
1880 K. Moeller, Brachwede Germany. Applied Rafinesque's (1819) theory of electrical precipitations of atmospheric dust but not successfully. (7)
1880 James G. Garland, Biddeford, Me. Invented "Apparatus for Moistening Air," a combination of air and water jet humidifying head. Filed Oct. 18, 1880. U.S. Patent #236,319, issued Jan. 4, 1881.
1881 August Kind, Berlin, Germany. Invented hydraulic ventilator to moisten air. U.S. Patent #264,149. Issued August 23, 1881.

1880 Robert Portner and B. Edward J. Eils, Alexandria, Virginia. Invented "Process of and Apparatus for Cooling Air." Blew air over refrigerated pipes, specified operation to prevent accumulation of "snow" on coils. System commended in paper published by U.S. Brewers' Association, New York, May 19, 1885. (26)
1880 A man in Staten Island, New York. Cooled restaurant by blowing air through pipes imbedded in ice and salt. (16)
1880 Madison Square Theatre, New York. Put ice in air stream (4 tons per evening) of B. F. Sturtevant Company fan-heating system to cool theater. (3)

1881 Fan and Ice Cooling System. U.S. White House, Washington, D.C. Room occupied by President Garfield during his illness July and

RELATED DEVELOPMENTS AND THEORIES	HEATING AND VENTILATING FANS, HEATERS, HEAT PUMPS	REFRIGERATION (NOT INCLUDING COMMERCIAL EQUIPMENT AS DOMESTIC REFRIGERATORS, ETC.)	AIR CLEANING, HUMIDIFYING, PURIFYING. WASHERS, HUMIDIFYING UNITS OR HEADS, ODOR ABSORBERS, ETC.	AIR COOLING. FAN-ICE, COOLING-COIL AND FAN-COOLING-COIL UNITS AND SYSTEMS	AIR CONDITIONING. SPRAY TYPE CENTRAL STATION APPARATUS AND CHEMICAL DEHUMIDIFIERS
YEAR	YEAR	YEAR	YEAR	YEAR	YEAR
			1881 William V. Wallace, Dorset, Vermont. Invented "Air Cooling Fountain and Air Cooling Apparatus"—a portable room humidifier. Filed April 8, 1881. U.S. Patent #242,083. Issued May 24, 1881.	August, 1881 cooled by melting about 436 lbs. of ice an hour. Report of Officers of the Navy on Ventilating and Cooling Executive Mansion. Vol. 8, Government Printing Office. (3)	
1882 Professor Ernest Mueller, Hanover, Germany. Reported on relation between regain, atmospheric moisture, and air temperature, and textile fibers. (24)					
1883 A. M. Butz. Invented a thermostatic control instrument. Draft regulator operated electrically, consisted of strip of metal fastened to strip of rubber. (27)			1883 George D. Bancroft, Springfield, Massachusetts. Invented moistening or humidifying head. U.S. Patent #287,898. Issued November 6, 1883. 1883 Frederic Tudor, Boston, Massachusetts. Devised humidifying system for Metropolitan Opera House, New York. Put pans of water containing heating coil under seat with up-discharge outlets. (3) Obtained two U.S. Patents:	1883 Restaurant at the Hygiene Exhibition, Berlin, cooled by blowing air over ice. (16)	

(1) "Steam and Hot Water Apparatus" #278,636. Issued May 29, 1883.

(2) "Apparatus for Heating by Exhaust Steam" #283,537. Issued August 21, 1883.

1883 Franz Windhausen, Berlin, Germany. Invented "Apparatus for Purifying Air and Gases." Filed Sept. 14, 1883. U.S. Patent #306,040. Issued September 30, 1884.

1884 Sir Oliver Lodge (1851-1940) British Physicist. Carried on extensive research on electric precipitation. He and associates applied process to lead smelters. (7)

1884 The Carbondale Machine Company, Carbondale, Pennsylvania (now Worthington Pump and Machinery Company, Harrison, New Jersey). Imported "Pontifex" absorption refrigerating machine from England, improved design and began manufacturing it in 1884. First machines were "live steam" absorption type. (28)

1884 George M. Capell and G. S. Ma-Beam. Invented "The Capell" centrifugal fan. Two rotors, one within the other, all blades backward curved. (18) U.S. Patent #291,493. Issued January 8, 1884.

1884 Franz Windhausen, Berlin, Germany. Invented "Refrigerating Rooms and Liquids and Apparatus Used Therefor." Machine chilled brine for cooling coils in air circulating system. Filed November 10, 1884. U.S. Patent #323,767. Issued August 4, 1885.

1884 William V. Wallace, Boston, Massachusetts. Invented "Air Cooling Device" —air passed over wetted disc for evaporative cooling. Filed March 4, 1884, U.S. Patent #297,476. Issued April 22, 1884. Also circulated air through chamber with jet of water discharged into air stream. Filed March 4, 1884, U.S. Patent #297,039. Issued April 15, 1884.

RELATED DEVELOPMENTS AND THEORIES	HEATING AND VENTILATING FANS, HEATERS, HEAT PUMPS	REFRIGERATION (NOT INCLUDING COMMERCIAL EQUIPMENT AS DOMESTIC REFRIGERATORS, ETC.)	AIR CLEANING, HUMIDIFYING, PURIFYING. WASHERS, HUMIDIFYING HEADS, ODOR ABSORBERS, ETC.	AIR COOLING. FAN-ICE, COOLING-COIL AND FAN-COOLING-COIL UNITS AND SYSTEMS	AIR CONDITIONING. SPRAY TYPE CENTRAL STATION APPARATUS AND CHEMICAL DEHUMIDIFIERS
YEAR	YEAR	YEAR	YEAR	YEAR	YEAR
1885 Professor Warren S. Johnson. Credited with inventing first automatic temperature control. (See 1883 above.) Led to formation of Johnson Service Company, Milwaukee, Wisconsin, in 1885. First control was an electric thermostat. In 1893 produced pneumatic thermostat. (29) 1886 Professor William Ferrel (1817–1891) American Meteorologist, U.S. Weather Bureau. Deduced empirical psychrometric formulae used as bases for U.S. Weather Bureau psychrometric tables. (31)		1887 The Creamery Package Mfg. Company, Chicago, Illinois. Established in 1887.	1885 Mortimer Sherman, Lowell, Massachusetts. Invented "Apparatus for Moistening Air in Cotton Mills, etc." U.S. Patent #324,043. Issued August 11, 1885. 1887 B. F. Sturtevant Company, Boston, Massachusetts. Installed heating, ventilating, and moistening system in Pacific Mills. Moistening means was a "simple rose nozzle" located in main duct at "mouth of fan." In 1889 similar system installed in Globe Yarn Mill at Fall River, Mass. (30)		

1888 Gotham Patented extended surface for heat transfer—helically wound finned tubing. Automotive industry employed surface for car radiators.

1889 The New York Blower Company (formerly New York), Chicago. Founded in 1889. Named its first centrifugal fan "Seri-Vane" in honor of Prof. Ser who in 1862 developed laws for fan design. (34)

1888 Representative of four refrigerating machine manufacturers met to consider standardization of products: Frick Company, De La Vergne, Consolidated, and National. Fifteen years lapsed before further cooperative efforts were attempted. (32)

1888 American Moistening Company, Providence, Rhode Island (Founded 1888) Sold humidifying heads, invented by Emil H. C. Oehlman in 1880 (see above) for rayon winding room. (First U.S. Patent for processing nitrocellulose rayon granted in 1884.) (33)

1888 Dr. Kilvington, Pres. Board of Health, Minneapolis, recommended fire to destroy odors. (22)

1888–1889 Walter B. Snow, B. F. Sturtevant Co., Boston, Massachusetts. Cited comparative costs of moistening air in textile mills as a function of the heating and ventilating system and as separate operation with direct heating system. (35)

1888 The Carbondale Machine Company, Carbondale, Pennsylvania. (See column 3, 1844.) Installed its first of many systems to cool air for industrial processes. First installation, coils on walls of plant of Hawley & Hoops Company, New York, 10-ton capacity.

RELATED DEVELOPMENTS AND THEORIES	HEATING AND VENTILATING FANS, HEATERS, HEAT PUMPS	REFRIGERATION (NOT INCLUDING COMMERCIAL EQUIPMENT AS DOMESTIC REFRIGERATORS, ETC.)	AIR CLEANING, HUMIDIFYING, PURIFYING. WASHERS, HUMIDIFYING UNITS OR HEADS, ODOR ABSORBERS, ETC.	AIR COOLING. FAN-ICE, COOLING-COIL AND FAN-COOLING-COIL UNITS AND SYSTEMS	AIR CONDITIONING. SPRAY TYPE CENTRAL STATION APPARATUS AND CHEMICAL DEHUMIDIFIERS
YEAR	YEAR	YEAR	YEAR	YEAR	YEAR
1890 Charles Dudley Warner (1829–1900) American Editor. Wrote in *Courant*, Hartford, Connecticut, "Everyone is talking about the weather but *no* one is doing anything about it." Comment often erroneously attributed to Mark Twain.		1890 St. Louis Automatic Refrigerating Company. Installed a central plant refrigerating system to cool restaurants, beer halls, etc. Cooling coils placed on walls midway between floor and ceiling. (37) System still in operation in 1916. (38)		1890 (Circa) C. W. Vollmann, Co-worker of Dr. Carl Linde (see column 3, 1873), and first president of Linde Canadian Refrigeration, Montreal, Canada. Installed air cooling system for Indian Rajah's palace. Coils sprayed with brine for continuous defrosting. (36) 1890 Alfred R. Wolff (1859–1909) Consulting Engineer, New York. Included space for ice in outdoor air intake of heating and ventilating system for Carnegie Hall, New York. (3) (16) 1890 Lennox Lyceum Theater, New York. Used artesian water through heating coils to cool air. Fan-heating system by B. F. Sturtevant. Johnson Electric Heat Regulator on temperature control. (3)	

1892 Alfred R. Wolff (see column 5, 1890) Consulting Engineer, New York. Wrote "Artificial Cooling of Air for Ventilation" published in *Engineers News*, July 7, 1892. (19)

1893 Theophile Schoesing, Fils. Reported research on Hygroscopic properties of cotton, wool, and silk. (24)

1893 Alfred R. Wolff, Consulting Engineer, New York. Introduced heat-unit practice in heating calculations. Introduced thermostatic control for heating systems in high class residences. (40)

1891 (Circa) B. F. Sturtevant Company, Boston, Mass. Published its first edition of *Ventilation and Heating* (from 4th edition, copyrighted 1904).

1891 *Ice and Refrigeration*, Chicago, Illinois. Published first issue of the magazine July 1, 1891. (9)

1892 Camille Edmond Auguste Rateau, St. Etienne, France. Invented multi-blade, high efficiency centrifugal fan. Filed September 2, 1892. U.S. Patent #500,080. Issued June 30, 1893.

1892 De La Vergne Refrigerating Machine Company, New York. Built the then largest refrigerating machine—ammonia reciprocating machine, 500 tons capacity. Installed in the Anheuser-Busch Brewing Association plant, St. Louis, Missouri, for wort cooling. (39)

1891 Eastman Kodak Company, Rochester, New York. Installed cooling coils on walls of film and plate coating rooms. De La Vergne's 50 ton ammonia machine supplied refrigeration. (System not installed for removing moisture from air but dehumidification occurred.) (38)

1892 Broadway Theater New York. Cooled by ice placed in a fan-ventilating system. (42)

1892 Private home, Frankfurt, Germany. Refrigerated coils in attic cooled air, delivered by gravity through ducts to rooms below. (16) (41)

1893 Blocks of ice placed in "Airways," of ventilating system to cool House of Lords and House of Commons, London. (3)

1893 Professor R. Ogden Doremus, College of City of New York. Wrote in *North American Review*, May, 1893. "If they can cool dead hogs in Chicago why not live bulls and bears in the New York Stock Exchange."

RELATED DEVELOPMENTS AND THEORIES	HEATING AND VENTILATING FANS, HEATERS, HEAT PUMPS	REFRIGERATION (NOT INCLUDING COMMERCIAL EQUIPMENT AS DOMESTIC REFRIGERATORS, ETC.)	AIR CLEANING, HUMIDIFYING, PURIFYING. WASHERS, HUMIDIFYING UNITS OR HEADS, ODOR ABSORBERS, ETC.	AIR COOLING. FAN-ICE, COOLING-COIL AND FAN-COOLING-COIL UNITS AND SYSTEMS	AIR CONDITIONING. SPRAY TYPE CENTRAL STATION APPARATUS AND CHEMICAL DEHUMIDIFIERS
YEAR	YEAR	YEAR	YEAR	YEAR	YEAR
				1894 The Creamery Package Mfg. Co., Chicago, Illinois. Installed cooling system for butter making. (Candy making in 1896.)	
1895 American Society of Heating and Ventilating Engineers founded in 1895. 1895 William D. Hartshorne, Lawrence, Mass. Began tests on moisture regain of textiles. (24) 1895 Prof. De Volson Wood, Stevens Institute of Technology. Employed "BTU" for British thermal units in *Thermodynamics* written by him and published by John Wiley & Sons, New York, 1895. Unit employed as early as 1851 although not named in Appleton's Dictionary of Mechanics in 1851. William John Macquorn Rankine in "Steam Engines" published in 1885 used "unit of heat" and "thermal unit." By 1897 term was accepted by engineers. (45)	1895 Professor Rolla C. Carpenter, Cornell University, Ithaca, New York. Wrote *Heating and Ventilating Buildings, an Elementary Treatise.* Published by John Wiley & Sons, New York and Chapman & Hall, Ltd., London, 1895. 1895 M. C. Huyett, Chicago, Illinois. Wrote *Mechanical Heating and Ventilation.* Published by The Henry O. Shepard Company, Chicago.	1895 De La Vergne Refrigerating Machine Company, New York. Installed ice-making plant for skating rink at New York Hippodrome, later refrigeration system was used for air cooling. (See Walter L. Fleisher, column 4, 1905.)	1895 Stuart W. Cramer (1868–1950) Textile Mill Engineer. Became Southern Agent for Whitin Machine Works, (43) and started outstanding work which later had large part in air conditioning industry.	1895 Linde Canadian Refrigeration Co. Ltd., Montreal. Installed Fan-coil (with brine spray defrosting) for air cooling: Gould Cold Storage Co., Montreal (System still in operation in 1950). In 1896, Lovell & Christmas, Ltd., Montreal, cold storage. System in part still in operation in 1950.	

1896 Fournier and Cornu Fans. Patented in France in 1896. Never filed in United States. Fans were forerunners of modern narrow blade multivane centrifugal fans. (18)

1896 Buffalo Forge Co., Buffalo, New York. Installed fan-ice air-cooling and ventilating system in Auditorium Hotel, Chicago. (44)

1897 Prof. R. C. Carpenter and W. G. Walker wrote informative article on theory and practice in design of fans and blowers, published in *Heating and Ventilating*, December 1897. (19)

1897 Elbert A. Corbin, Philadelphia, and Charles E. Foster, Washington, D.C. Invented a "Heating and Ventilating Device." Disc fan in outside wall opening located back of floor type radiator. U.S. Patent #593,737, issued Mar. 16, 1897.

1897 Kroeschell Bros. Ice Machine Company, Chicago. Established 1897. Built carbon dioxide reciprocating machines. (46) In 1922 joined Brunswick Refrigerating Company, New Brunswick, N.J. In 1930 became part of Carrier Corporation, Syracuse, N.Y.

1897 Joseph McCreery, Toledo, Ohio. Invented apparatus in which air was drawn over tank of water for cleaning and compartment filled with ice for cooling. Filed February 17, 1897, U.S. Patent #586,365. Issued July 13, 1897. Invented "Air Cleaning and Cooling Device." Tiers of pans from which water spilled into path of air stream. Filed November 5, 1897, U.S. Patent #626,388. Issued June 6, 1899.

1897 Joseph McCreery, Toledo, Ohio. Added ice compartment to Air Cooling, Cleansing and Ventilating Device. (See column 4, 1897.)

1897 Kroeschell Bros. Ice Machine Company, Chicago. Installed refrigerated coils on chocolate dipping room walls in candy factory of Jos. B. Funke Co., La Crosse, Wisconsin. (47)

1897 Edwin F. Portor, (Bay State Electric Heat and Light Co., Jersey City, N.J.). Invented "Apparatus for Cooling and Agitating Air," a disc-type-fan cold diffuser. Filed December 20, 1897, U.S. Patent #702,994. Issued June 24, 1902.

RELATED DEVELOPMENTS AND THEORIES	HEATING AND VENTILATING FANS, HEATERS, HEAT PUMPS	REFRIGERATION (NOT INCLUDING COMMERCIAL EQUIPMENT AS DOMESTIC REFRIGERATORS, ETC.)	AIR CLEANING, HUMIDIFYING, PURIFYING. WASHERS, HUMIDIFYING UNITS OR HEADS, ODOR ABSORBERS, ETC.	AIR COOLING. FAN-ICE, COOLING-COIL AND FAN-COOLING-COIL UNITS AND SYSTEMS	AIR CONDITIONING. SPRAY TYPE CENTRAL STATION APPARATUS AND CHEMICAL DEHUMIDIFIERS
YEAR	YEAR	YEAR	YEAR	YEAR	YEAR
1898 Walter B. Snow, B. F. Sturtevant Company, Boston, Mass. Edited *Mechanical Draft, a Practical Treatise.* Published by B. F. Sturtevant Co. Edited second book on heating and ventilating same year. Both advanced the art of fan engineering.	1898 Samuel Cleland Davidson, Belfast, Ireland. Invented "Sirocco" Fan. Design set new standards for centrifugal fan performance. British Patent #4609. Issued 1898. American Blower Corporation, Detroit, acquired rights in 1909. 1898 Henry Baetz Invented air heating apparatus. Disc-fan-type unit heater built by St. Louis Blower and Heater Company, St. Louis, U.S. Patent issued May 17, 1894. (48)	1899 The Carbondale Machine Company, Carbondale, Pa. Supplied refrigerating system to cool autopsy room, Cornell Medical College, New York (see column 5, 1899).	1899 John Zellweger, St. Louis, Missouri. Invented "Apparatus for Collecting Gases, Vapors, and Foreign Particles from Air." Filed September 26, 1899. U.S. Patent #679,587. Issued July 30, 1901. Vertical unit, disc fan, water flowing downward. 1899 B. F. Sturtevant Co., Boston, Mass. Installed "Heating Apparatus" for Hamilton-Brown Shoe	1899 Alfred R. Wolff, Consulting Engineer, New York. Designed fan-coil air cooling system for autopsy room, Cornell Medical College, New York. The Carbondale Machine Company supplied 50 ton capacity refrigerating system. (16)	

1899 (or earlier) Canadian Ice Machine Company, Toronto, representative of York Corporation, York, Pa. Installed air cooling fan-coil system in Casa Loma Castle to cool wine cellar and portions of the castle.

1900 Court Theater, Vienna. Underground air passage of ventilating system divided, one included water spray for evaporative cooling. System operated on by-pass principle. (16) (41)

1900 (Circa) Samuel Cleland Davidson, Belfast, Ireland. Included ice for summer air cooling in combination with air washer and filter. (See column 4, 1900.)

Co., St. Louis, Mo. A fan-coil indirect heating system with an air washer that included "wash tubes" after spray chamber. A similar system was installed for U.S. Playing Card Co., Norwood, Ohio the same year.

1900 Richard H. Thomas, Chicago. Invented "Air Purifying Apparatus." Horizontal air washer with vertical banks of spray nozzles. Unit became the "Acme" washer. Filed April 20, 1900, U.S. Patent #655,285, issued August 7, 1900.

1900 (Circa) Samuel Cleland Davidson, Belfast, Ireland. Designed combination air washer and filter consisting of cylindrical drum, lower segment rotated in tank of water. Apparatus included ice for cooling air in summer.

1900 B. F. Sturtevant Co., Boston, Mass. Installed "Heating and Ventilating Apparatus" for William Building, Cleveland, Ohio. System included an air washer similar to one used on 1899 installations (see B. F. Sturtevant above, 1899).

1900 Twenty-six refrigerating machine companies advertised in *Ice and Refrigeration,* July 1900. In 1952 five are in business under same name:

Creamery Package Mfg. Co.; Frick Co.; J. & E. Hall, Ltd., London; Vilter Manufacturing Co.; and Henry Vogt Machine Co.

1900 Brunswick Refrigerating Company, New Brunswick, N.J. Founded in 1900. 1922 combined with Kroeschell Bros. Ice Machine Co., Chicago to form Brunswick-Kroeschell Co. In 1930 became part of Carrier Corporation, Syracuse, N.Y.

1900 Soren Thurstensen, Louisville, Ky. Invented "Ammonia Still." Assigned to Henry Vogt Machine Co., Louisville, Ky. Filed August 2, 1900, U.S. Patent #671,810. Issued April 9, 1901.

1900 Alexander Robinson and Kenneth Cameron McDowell. Invented "Improvement in Apparatus for Heating Air." Applicable for drying yarns and fabrics. Pipe coil heater and blow-through disc fan. U.S. Patent #22,793, issued May 7, 1900.

1900 Prof. C. F. Marvin, U.S. Weather Bureau, Washington, D.C. Devised "Psychrometric Tables for Obtaining the Vapor Pressure Relative Humidity and Temperature and the Dew Point of Air." (20)

1900 Frederick R. Still and Charles H. Treat, American Blower Corporation, Detroit. Perfected design and improved accuracy of Pitot tube.

RELATED DEVELOPMENTS AND THEORIES	HEATING AND VENTILATING FANS, HEATERS, HEAT PUMPS	REFRIGERATION (NOT INCLUDING COMMERCIAL EQUIPMENT AS DOMESTIC REFRIGERATORS, ETC.)	AIR CLEANING, HUMIDIFYING, PURIFYING. WASHERS, HUMIDIFYING UNITS OR HEADS, ODOR ABSORBERS, ETC.	AIR COOLING. FAN-ICE, COOLING-COIL AND FAN-COOLING-COIL UNITS AND SYSTEMS	AIR CONDITIONING. SPRAY TYPE CENTRAL STATION APPARATUS AND CHEMICAL DEHUMIDIFIERS
YEAR	YEAR	YEAR	YEAR	YEAR	YEAR
1902 Willis H. Carrier, Buffalo Forge Company, Buffalo, New York. Conceived idea for control of moisture in air from observation of a fog in Pittsburgh. Idea developed as dew-point control, the basis for all central station spray type air-conditioning systems. 1902 Buffalo Forge Company, Buffalo, N.Y. Formed "Department of Experimental Engineering," first research laboratory in fan-heating and air conditioning industry (51) (at Willis H. Carrier's suggestion).	1902 Willis H. Carrier (1876–1950) Buffalo Forge Company, Buffalo, N.Y. Set up test to obtain heat transfer data on pipe coil heaters. (50) 1902 Frederick R. Still, American Blower Corporation, Detroit. Read paper before Western Society of Engineers that had "more real meat than one can find in many volumes by others on fans." (34)	1901 Brunswick Refrigerating Machine Company, New Brunswick, N.J. Produced a self-contained, 3-cylinder, radial ammonia condensing unit. Machine received "Highest Award" medal, Pan American Exposition, Buffalo, N.Y., 1901. 1902 De La Vergne Refrigerating Machine Company, New York. Supplied refrigerating system for air conditioning installation, Sackett-Wilhelms Lithographing and Printing Company, Brooklyn, N.Y. 1902 The Carbondale Machine Company, Carbondale, Pa. Supplied refrigerating systems for fan-coil air-cooling installations: (A) New York Stock Exchange, New York; (B) Hanover National Bank, New York. (16) (41) (49) 1902 Charles A. Parsons, New Castle-upon-Tyne, England. Received first of his patents on steam turbines. Worked at this	1902 John William Fries, Salem, N.C. Invented "Apparatus for Humidifying and Cleaning Air." Filed April 24, 1902, U.S. Patent #831,275, issued September 18, 1906.	1901 Scranton High School, Scranton, Pa. Ice put into outdoor air intake of fan-ventilating system to cool air in auditorium during commencement exercises. Procedure repeated in 1902. Eight tons of ice used per evening. (42) 1902 Alfred R. Wolff, New York Consulting Engineer. Designed fan-coil cooling and ventilating system for New York Stock Exchange. System operated satisfactorily for twenty years. The Carbondale Machine Co., Carbondale, Pa. supplied 3 100-ton capacity refrigerating systems. (49) 1902 Alfred R. Wolff, New York Consulting Engineer. Designed fan-coil cooling and ventilating system for Hanover National Bank, New York. The Carbondale Machine Company supplied 1 100-ton-capacity refrigerating machine. (16)	1902 Willis H. Carrier, Buffalo Forge Company, Buffalo, N.Y. Designed fan-coil dehumidifying system for Sackett-Wilhelms Lithographing and Printing Company, Brooklyn, N.Y. Installation differed from other fan-coil systems as design conditions were based on control of moisture in air. Walter W. Timmis, New York, Consulting Engineer, J. Irvine Lyle, Buffalo Forge Co., Sales Engineer, De La Vergne Refrigeration Machine Co., New York furnished 30-ton refrigerating system. Johnson Service Co., Milwaukee furnished thermostat used on

mixing dampers of a double duct system and humidistats on supply of steam to perforated pipes in risers which humidified the air.

1903 Thomas Chester, Davidson & Company, Ltd. Designed forced air heating humidifying, ventilating system, including ice for air cooling, for Robinson & Cleaver, Ltd., London (Department Store). Employed Sirocco fan and washer invented by Samuel Cleland Davidson, Belfast, Ireland in 1900.

1903 Kroeschell Ice Machine Company, Chicago. Installed six coil-cooling systems for chocolate dipping room for National Candy Co. Coils built into sheet metal housing unit type with CO_2 refrigerant.

1903 Louis Herman Brinkman, Whitlock Coil Pipe Co., West Hartford, Conn. Invented "Cooler," disc-type-fan cold diffuser. Filed January 29, 1903, U.S. Patent #843,864, issued February 12, 1907.

1903 Alfred R. Wolff, New York, Consulting Engineer. Designed forced air heating, ventilating and humidifying system for residence at 81 St. and 5th Ave., New York. Walter L. Fleisher (Francis Brothers and Jellet Company) installed system.

time on steam jet refrigeration, independently and simultaneously with Maurice Leblanc, of Paris, France (see below 1905).

1903 John J. Spear, Wilmette, Ill. Invented new type of heater. Trade name Vento. Rapidly replaced prime surface pipe coil in fan-heating installations. American Radiator Company, Chicago, placed Vento on market early in 1905. Used extensively until extended surface tubes were developed for air heating in 1922.

1903 Maurice Leblanc, Paris, France, Westinghouse Co. Invented "Refrigerating Machine." Employed centrifugal compression using water vapor as the refrigerant. France Patent issued October 27, 1903. Filed February 24, 1904, U.S. Patent #977,659, issued December 6, 1910. Machine never got beyond laboratory stage. (52)

1903 Actual start of Refrigerating Manufacturers Association after original attempt to form organization in 1888 did not succeed.

1903 Kroeschell Ice Machine Company, Chicago. (See column 5)

RELATED DEVELOPMENTS AND THEORIES	HEATING AND VENTILATING FANS, HEATERS, HEAT PUMPS	REFRIGERATION (NOT INCLUDING COMMERCIAL EQUIPMENT AS DOMESTIC REFRIGERATORS, ETC.)	AIR CLEANING, HUMIDIFYING, PURIFYING. WASHERS, HUMIDIFYING UNITS OR HEADS, ODOR ABSORBERS, ETC.)	AIR COOLING. FAN-ICE, COOLING-COIL AND FAN-COOLING-COIL UNITS AND SYSTEMS	AIR CONDITIONING. SPRAY TYPE CENTRAL STATION APPARATUS AND CHEMICAL DEHUMIDIFIERS
YEAR	YEAR	YEAR	YEAR	YEAR	YEAR
1904 *Heating and Ventilating.* Published its first issue June 1904. 1904 Alfred R. Wolff, New York, Consulting Engineer. He was referred to as the "leading factor in the heating and ventilating field in this country." (53) 1904 Stuart W. Cramer, Charlotte, N.C. Invented heat and humidity control instruments, electrically operated. (A) Filed December 28, 1904. U.S. Patent #811,833, issued January 30, 1906. (B) Filed December 29, 1904. U.S. Patent #813,883, issued February 20, 1906. (Instruments put on market in 1907.) 1904 B. F. Sturtevant Company, Boston, Mass. Initiated research in 1904. (51)	1904 Hugo Junkers, Aix-la-Chapelle, Germany. Invented "Heating and Cooling Apparatus," disc-fan-type unit heater or cooler. Filed November 8, 1904. U.S. Patent #1,208,790, issued December 19, 1916.	1904 The American Society of Refrigerating Engineers. Founded in 1904. 1904 Thomas Shipley, York Corporation, York, Pa. Published data from tests made by his company, to standardize ton of refrigeration in Btu. (54) 1904 York Corporation, York, Pa. Supplied refrigerating system for James Gayley system to freeze moisture out of blast air for Carnegie Steel Co., Etna, Pa. (56)	1904 Willis H. Carrier, Buffalo Forge Co., Buffalo, N.Y. Treating Air" invention installed as air washer in fan-heating system in La Crosse National Bank, La Crosse, Wisconsin.	1904 John Zellweger, St. Louis, Mo. Invented "Air Filter and Cooler." Centrifugal fan, outer casing of porous material and means for keeping material wetted for evaporative cooling. Filed August 13, 1904. U.S. Patent #789,247, issued May 9, 1905. Advantage of equipment was space saving. (55) 1904 York Corporation, York, Pa. Installed fan-coil cooling system in Kolbs Bakery, Philadelphia. 1904 Maurice Leblanc, Paris, France. Westinghouse Co. Invented "Cooling Apparatus for Houses." Proposed fountains of water cooled by steam jet refrigeration be installed in rooms. No forced air circulation. Filed February 10, 1904. U.S. Patent #867,024, issued August 9, 1910.	1904 Willis H. Carrier, Buffalo Forge Co., Buffalo, N.Y. Invented "Apparatus for Treating Air." Equipment to saturate air at specified temperature. Filed September 16, 1904. U.S. Patent #808,897, issued August 9, 1910. 1904 James Gayley (1855–1920). Devised system to freeze moisture out of blast furnace air for Carnegie Steel Co., Etna, Pa. York Corporation, York, Pa. supplied 2 225-ton ammonia refrigerating systems. (56)

1904 Hugo Junkers, Aix-la-Chapelle, Germany. Invented heating and cooling apparatus (see column 2, 1904).

1905 Stuart W. Cramer, Charlotte, N.C. Invented "Automatic Indicating and Regulating Hygrometer." Filed January 9, 1905. U.S. Patent #809,672, issued January 9, 1906. This was "Type ER Regulator."

1905 John H. Kinealy, Ferguson, Missouri. Wrote *Centrifugal Fans*. Published by Spon and Chamberlair, New York, N.Y.

1905 Maurice Leblanc, Paris, France. Westinghouse Co. Invented Steam Ejector Refrigerating Apparatus. Filed February 15, 1905. U.S. Patent #1,005,851, issued October 17, 1911.

1905 Stuart W. Cramer, Charlotte, N.C. Invented "Humidifying and Ventilating Apparatus." Filed June 1, 1905. U.S. Patent #821,989, issued May 29, 1906.

1905 Joseph J. Smith, New York. Invented "Humidifier." U.S. Patent #803,022, issued October 31, 1905.

1905 John H. Kinealy, Ferguson, Missouri. Invented "Air-purifying Apparatus." (Later became the "Webster Air Washer.") Filed March 17, 1905. U.S. Patent #813,217, issued February 20, 1906. Second U.S. Patent issued February 20, 1906.

1905 Jean Hasé, Wiesbaden, Germany. Invented "Humidifier." U.S. Patent #804,120, issued November 7, 1905.

1905 Walter L. Fleisher (Francis Brothers and Jellet Co.) New York. Installed fan-coil cooling and ventilating system at New York Hippodrome, New York. Refrigerating Machine for Ice Rink (see column 3, 1895), supplied refrigeration for cooling coil—100-ton-capacity system. (57)

1905 Fred Wittenmeir, Kroeschell Bros. Ice Machine Company, Chicago. Designed direct expansion air cooler used in connection with air-washing system. (47) (First used dry coil surface cooling, later used sprays to wet coil surface for increased heat transfer.)

1905 Richard H. Thomas, Chicago. Invented "Humidity Reducing Apparatus." Added to his air washer a drying chamber in which a drying agent was discharged. Drying agent not named in patent. Filed October 23, 1905. U.S. Patent #871,194, issued November 19, 1907.

1905 Guiseppe Cattaneo of Charlottenburg and Julius Schlesinger of Berlin, Germany. Invented "Apparatus for Drying Air for Metallurgical Purposes." Filed January 9, 1905. U.S. Patent #839,362, issued December 25, 1906.

1905 Eastman Kodak Company, Rochester, N.Y. Installed direct expansion ammonia refrigerating coil in tunnel to dehumidify 15,000 cfm of air—film-festooned-type dryer. Interchanger or economizer employed to reheat dehumidified air.

RELATED DEVELOPMENTS AND THEORIES	HEATING AND VENTILATING FANS, HEATERS, HEAT PUMPS	REFRIGERATION (NOT INCLUDING COMMERCIAL EQUIPMENT AS DOMESTIC REFRIGERATORS, ETC.)	AIR CLEANING, HUMIDIFYING, PURIFYING. WASHERS, HUMIDIFYING UNITS OR HEADS, ODOR ABSORBERS, ETC.	AIR COOLING. FAN-ICE, COOLING-COIL AND FAN-COOLING-COIL UNITS AND SYSTEMS	AIR CONDITIONING. SPRAY TYPE CENTRAL STATION APPARATUS AND CHEMICAL DEHUMIDIFIERS
YEAR	YEAR	YEAR	YEAR	YEAR	YEAR
	1906 Ilg Electric Ventilating Company, Chicago. Incorporated October, 1906. (Entered unitary air washing, humidifying, and cooling market in 1917.)		1906 Stuart W. Cramer, Charlotte, N.C. Invented two types of humidifying heads, the second installed in many textile mills. (A) Filed March 14, 1906. U.S. Patent #840,917, issued January 8, 1907. (B) Filed April 18, 1906. U.S. Patent #852,823, issued May 7, 1907. 1906 John W. Fries. Invented "Air Humidifying and Cleaning Apparatus." U.S. Patent #831,275, issued September 18, 1906. Unit later improved by F. F. Bahnson, Winston-Salem, N.C. and sold to textile mills and tobacco plants in 1908. (The Bahnson Company, Winston-Salem, N.C. established in 1915.) 1906 B. F. Sturtevant Company, Boston, installed heating, ventilating, air washing and coil cooling system for Battle Mountain Sanatorium. (See column 6, 1911.)	1906 American Blower Corporation, Detroit. Began installing fan-coil cooling, heating, ventilating systems. 1906 York Corporation, York, Pa. Installed fan-coil cooling system for Ward Corby Company, Cambridge, Mass. 1906 Arnold H. Goelz, Kroeschell Bros. Ice Machine Co., Chicago. Designed CO_2 refrigerating machine, 150 tons, and coils for air-cooling system for the Pompeiian Room, Congress Hotel, Chicago. (Spray-type air conditioning apparatus added in 1911.) (41) 1906 A. M. Feldman, New York Consulting Engineer. Designed fan brine-coil cooling, ventilating, and heating system for Kuhn Loeb & Co. bank building. Air washer operated for winter humidifying only. (58) 1906 B. F. Sturtevant Co., Boston, Mass. Installed "Heating and Ventilating Apparatus," for Battle	1906 Stuart W. Cramer, Charlotte, N.C. Introduced term, Air Conditioning, as control of moisture in air, May 1906. (60) 1906 Willis H. Carrier, Buffalo Forge Company, Buffalo, N.Y. Invented what later became known as "Dew-Point Control," a basic patent, filed July 16, 1906. U.S. Patent #854,270, issued May 21, 1907. 1906 Carrier Psychrometric Chart published in Buffalo Forge Catalogue #97, copyright 1906. 1906 Buffalo Forge Company, Buffalo, N.Y. Installed Carrier designed central station spray-type air conditioning apparatus in connection with a fan-heating system installed by B. F. Sturtevant Company, Boston, in Chronicle Cotton Mill, Belmont, N.C.

Mountain Sanatorium, Hot Springs, South Dakota. System included refrigerated coil in series with heating coil with a by-pass. Drawings indicate a spray apparatus with by-pass was supplied by Buffalo Forge Company.

1906 B. F. Sturtevant Company, Boston, Mass. Installed fan-coil cooling system in Walter Baker & Co. Milton, Mass., plant. The Carbondale Machine Company furnished 3 100-ton ammonia absorption machines to chill brine coils. System still in operation in 1952.

1906 The Carbondale Machine Company, Carbondale, Pa. (See B. F. Sturtevant Company, column 5.)

1907 Parke-Davis & Company, Detroit, Michigan. Obtained permission to use drawing and specifications prepared by Willis H. Carrier and C. A. Booth, Buffalo Forge Company, to install year-round air conditioning system for capsule room. (Refrigeration recalled by Dr. Carrier as being supplied by Great Lakes Refrigerating Company, Detroit.) (61)

1907 Buffalo Forge Company, Buffalo, N.Y. Installed air-conditioning system with automatic humid-

1907 Walter L. Fleisher, New York. Designed air-handling system for Congress Hotel. Installation similar to Kuhn-Loeb & Co. building (See A. M. Feldman 1904 above). Refrigeration supplied by Kroeschell Bros. Ice Machine Co. (See 1906 above). Controls by Powers Regulator Company, Chicago. (57) (58) Installation later revised and B. F. Sturtevant Co. fan, heating coils, and Acme (Thomas) washer were installed. Washer was not operated during cooling

1907 Frederick Gardner Cottrell (1877-1948). University of California, Berkeley, Calif. Began experiments on deposit of smoke and dust by electrical precipitation. Invented electrical precipitator. Filed July 9, 1907. U.S. Patent #895,729, issued August 11, 1908. First commercial application of Cottrell Precipitator installed at Shelby Smelting and Lead Co. in California, in 1907. (59)

1907 Albert W. Thompson (1874-1940) Manchester, N.H. In-

1907 Willis H. Carrier, Buffalo Forge Company, Buffalo, N.Y. Invented two types of differential control instruments.
(A) Filed October 11, 1907, U.S. Patent #902,713, issued Nov. 1908.
(B) Filed October 11, 1907, U.S. Patent #896,690, issued Aug. 18, 1908.
(These were first of many patents on controls issued to Willis H. Carrier.)

RELATED DEVELOPMENTS AND THEORIES	HEATING AND VENTILATING FANS, HEATERS, HEAT PUMPS	REFRIGERATION (NOT INCLUDING COMMERCIAL EQUIPMENT AS DOMESTIC REFRIGERATORS, ETC.)	AIR CLEANING, HUMIDIFYING, PURIFYING, WASHERS, HUMIDIFYING UNITS OR HEADS, ODOR ABSORBERS, ETC.	AIR COOLING. FAN-ICE, COOLING-COIL AND FAN-COOLING-COIL UNITS AND SYSTEMS	AIR CONDITIONING. SPRAY TYPE CENTRAL STATION APPARATUS AND CHEMICAL DEHUMIDIFIERS
YEAR	YEAR	YEAR	YEAR	YEAR	
			vented "Humidifying Apparatus." Installed equipment in Amoskeag Mills, Amoskeag, N.Y. Controlled system with Carrier Differential Humidistat. Humidifiers later became known as "Turbo-Humidifiers" manufactured by Parks-Cramer Company, Fitchburg, Mass., and Charlotte, N.C. Filed Aug. 13, 1907. U.S. Patent #869,945, issued Nov. 5, 1907.	cycle as wetted coils were believed to "scrub" air. (58)	ity and dew-point control in Huguet Silk Mill, Wayland, N.Y. System designed by Willis H. Carrier, erected by E. P. Heckel, controls in charge of A. E. Stacey, Jr. Contract sold by W. S. Koithan. System is still in operation, **1952.**
			1907 (Circa) Stuart W. Cramer, Charlotte, N.C. Installed humidifying heads and automatic controls invented by him in Cramerton Mills, Cramerton, N.C.		1907 Alfred R. Wolff, New York, Consulting Engineer. Designed year-round air-conditioning system for Metropolitan Museum of Art, New York. (16)
			1907 H. Bentz Invented air-moistening apparatus. Blew air downward through pan of water. Two U.S. Patents #848,340 and #848,341 issued March 26, 1907.		1907 Warren Webster and Company, Camden, N.J. (Established 1888). Entered air conditioning industry.
					1907 William G. R. Braemer (1870–1944), (Formerly with Buffalo Forge Company). Joined Warren Webster and Company to head its air conditioning activities.
					1907 William G. R. Braemer, Camden, N.J. Invented a dew-point control arrange-

1908 Edward W. Comfort. Invented "Humidity Regulator." Filed March 14, 1908. U.S. Patent #977,933, issued Dec. 6, 1910. Assigned to Warren Webster and Company, Camden, N.J.

1908 Buffalo Forge Company, Buffalo, N.Y. Established an Instrument Department to build air conditioning control instruments. Employed Edward W. Comfort to take charge of department.

1908 Edward W. Comfort and Willis H. Carrier, Buffalo Forge Company. Invented Thermostatic "Regulating Device." Filed Sept. 28, 1908. U.S. Patent #929,655, issued Aug. 3, 1909.

1908 Prof. Arthur H. Baker. Installed radiant heating system in Liverpool Cathedral. Designated as "Romana" method of heating as it resembled old Roman heating practices. (62)

1908 Vilter Manufacturing Company, Milwaukee, Wisc. Supplied refrigerating system for dehumidifying blast air for Illinois Steel Company, South Chicago, Ill. (See column 6, 1908.)

1908 York Corporation, York, Pa. Supplied refrigerating system for air conditioning installation for Celluloid Corporation, Newark, N.J. (See column 6, 1908.)

1908 Kroeschell Bros. Ice Machine Company, Chicago. (See column 5.)

1908 Kroeschell Bros. Ice Machine Company, Chicago. Installed CO_2 fan-coil air-cooling system for smokeless powder room on U.S.S. Ohio.

1908 Willis H. Carrier, Buffalo Forge Company, Buffalo, N.Y. Invented spray apparatus with counter flow nozzles and flooded eliminator plates. Filed April 23, 1908. U.S. Patent #1,059,976, issued April 29, 1913.

1908 Carrier Air Conditioning of America, New York. Established by Buffalo Forge Company, Buffalo, as totally owned subsidiary to handle air conditioning sales and engineering.

1908 Harry Sloan, Vilter Manufacturing Company, Milwaukee, Wisc. Designed spray type dehumidifying system for blast furnace, Illinois Steel Co., South Chicago, Ill.

1908 Walter L. Fleisher, New York Consulting Engineer. Designed humidifying system for Owl Cigar Company, New York.

ment. Ingenious arrangement did not infringe on Willis H. Carrier's invention and gave almost as good results. Filed Sept. 11, 1907. U.S. Patent #885,173, issued April 21, 1908. (This was first of eleven or more patents on air conditioning issued to William G. R. Braemer.)

RELATED DEVELOPMENTS AND THEORIES	HEATING AND VENTILATING FANS, HEATERS, HEAT PUMPS	REFRIGERATION (NOT INCLUDING COMMERCIAL EQUIPMENT AS DOMESTIC REFRIGERATORS, ETC.)	AIR CLEANING, HUMIDIFYING, PURIFYING. WASHERS, HUMIDIFYING UNITS OR HEADS, ODOR ABSORBERS, ETC.	AIR COOLING. FAN-ICE, COOLING-COIL AND FAN-COOLING-COIL UNITS AND SYSTEMS	AIR CONDITIONING. SPRAY TYPE CENTRAL STATION APPARATUS AND CHEMICAL DEHUMIDIFIERS
YEAR	YEAR	YEAR	YEAR	YEAR	YEAR
1909 G. B. Wilson Wrote *Air Conditioning*—a treatise of textile mill humidification, ventilation, cooling, etc. Published by John Wiley & Sons, New York. Book contains an advertisement of Carrier Air Conditioning Company of America.					Perforated steam pipe in fan-heating ventilating system. (63) 1908 Willis H. Carrier, Buffalo Forge Company. Designed year-round air conditioning for Celluloid Corporation, Newark, N.J. York Corporation, York, Pa. supplied refrigerating system—125 tons capacity. (64) 1908 William G. R. Braemer, Warren Webster and Company, Camden, N.J. Designed spray-type central system for Armour & Co., Chicago, Office Building. Brine coil in tanks chilled spray water. 1908 Stuart W. Cramer, Charlotte, N.C. Installed air washer at Loray Mills, Gastonia, N.C., on existing B. F. Sturtevant Company fan-heating system. Used automatic humidity control invented by him.

mer, ... Charlotte, N.C. Wrote "Useful Information for Cotton Manufacturers," One volume devoted to discussion to humidifying and air conditioning textile mills.

1911 Willis H. Carrier, Buffalo Forge Company. Presented paper "Rational Psychrometric Formulae" before meeting of American Society of Mechanical Engineers. (68)

1911 Willis H. Carrier and Frank L. Busey, Buffalo Forge Company. Presented paper "Air Conditioning Apparatus" before meeting of American Society of Mechanical Engineers. (50)

... Soule, American Radiator Corporation, Buffalo, N.Y. Conducted tests on Vento heaters. Reported results at meeting of A.S.H.V.E. (65) Reported on further heater tests in 1913. (66)

1910 Domestic refrigerating machines were contemplated as early as 1910. J. M. Larson produced a manually operated domestic refrigerating machine in 1913. The Kelvinator Company sold the first domestic refrigerating machine in 1918. Designed by E. J. Copeland. (67)

Ice Machine Company, Chicago. (See column 5.)

1911 E. B. Freeman, B. F. Sturtevant Company, Boston. Invented railroad car cooling and ventilating system. (See column 5, 1911.)

Ice Machine Co., Chicago. Installed CO_2 coil-cooling system in office of Larkin Building, Buffalo, N.Y. Frank Lloyd Wright, Architect.

1910 Albert J. Dronsfield, Providence, R.I. Invented "Humidifying Apparatus," a spray nozzle for spray-type air washer. Filed Dec. 27, 1910. U.S. Patent #1,002,905, issued Sept. 12, 1911.

1911 Vilter Manufacturing Company, Milwaukee. Supplied refrigerating system for air conditioning installation for Mirror Candy Company, New York. (See column 6, 1911.)

1911 Henry Vogt Machine Company, Louisville, Ky. Supplied refrigerating system for air conditioning installation, Eli Lilly & Company, Indianapolis. (See column 6, 1911.)

1911 The Carbondale Machine Company, Carbondale, Pa. Supplied refrigerating system for Northern Iron Company, Stan-

Buffalo Forge Company, Buffalo, New York. Invented a humidifying head—the "Hygrogyre." Filed Dec. 31, 1909. U.S. Patent #971,248, issued Sept. 27, 1910. (Only a few of the units were built.)

1910 Charles H. Herter, Refrigerating Engineer, N.Y. Wrote article "Cooling Air in Summer," Southern Engineer, Oct. and Nov. 1910. Describes air cooling systems of that era.

1911 Air Cooler, invented by Frederick Wittenmeier, Kroeschell Bros. Ice Machine Co., Chicago. Installed air-cooling systems in Blackstone Hotel and Hotel Planters, Chicago, and Rogers Hotel, Minneapolis. (47) Filed May 18, 1911. U.S. Patent #1,003,129, issued Sept. 12, 1911. A cold coil unit in housing.

1911 E. B. Freeman, B. F. Sturtevant Company, Boston. Invented Pullman and private car cooling and ventilating system. Air cooled by being drawn around ice tanks.

mer and William B. Hodge, Charlotte, N.C. Invented central station spray-type air conditioning apparatus. Filed Dec. 4, 1909. U.S. Patent #960,830, issued June 7, 1910.

✓ 1910 Carrier Air Conditioning Company of America, New York. Installed year-round air conditioning system for Bosch Magneto Company, Springfield, Mass. Artesian well supplied cold water for dehumidifying air.

1911 Edward H. Vitalius, Detroit, Mich. Sold bakery air conditioning systems consisting of:
(1) American Blower Corporation, "Sirocco" fan.
(2) Carrier Air Conditioning Company, spray-type air conditioning apparatus.
(3) York Corporation, refrigerating system.

1911 Mirror Candy Company, New York. Installed Carrier Air Conditioning Company of America air conditioning system. Vilter Manufacturing Company supplied me-

137

RELATED DEVELOPMENTS AND THEORIES	HEATING AND VENTILATING FANS, HEATERS, HEAT PUMPS	REFRIGERATION (NOT INCLUDING COMMERCIAL EQUIPMENT AS DOMESTIC REFRIGERATORS, ETC.)	AIR CLEANING, HUMIDIFYING, PURIFYING. WASHERS, HUMIDIFYING UNITS OR HEADS, ODOR ABSORBERS, ETC.	AIR COOLING. FAN-ICE, COOLING-COIL AND FAN-COOLING-COIL UNITS AND SYSTEMS	AIR CONDITIONING. SPRAY TYPE CENTRAL STATION APPARATUS AND CHEMICAL DEHUMIDIFIERS
YEAR	YEAR	YEAR	YEAR	YEAR	YEAR
		dish, N.Y., blast furnace air conditioning. (See column 6, 1911.)			chanical refrigerating system. 1911 Eli Lilly & Company, Indianapolis, Indiana. Installed Carrier Air Conditioning Company of America air conditioning system for capsule room. Henry Vogt Machine Company, Louisville, Ky., supplied refrigerating system, 75-ton capacity. 1911 Walter L. Fleisher, New York, Consulting Engineer. Designed humidifying central station system with evaporative cooling for Folies Bergère (name later changed to Fulton) Theater, New York. Used Thomas "Acme" spray type apparatus. (63) 1911 Northern Iron Company, Standish, N.Y. Installed Carrier Air Conditioning Company of America air conditioning system to dehumidify blast furnace air. Automatic control for system invented by Edward T. Murphy (U.S. Patent #1,012,414). The Carbondale Machine Co., Carbondale, supplied refrigerating system

1912 J. Irvine Lyle, Carrier Air Conditioning Company of America, New York. Read paper "Relative Humidity—Its Effect on Comfort and Health" at A.S.H.V.E. annual meeting, 1912.

1912 Theory on axial-flow fans, later to fill a place in ventilating and air conditioning industry, was advanced by N. E. Joukovsky at Society of Mathematics, Moscow. (69)

1912 J. Irvine Lyle, Carrier Air Conditioning Company of America, New York. Delivered paper "Atmospheric Dehumidifying" before A.S.R.E., Dec. 2, 1912.

1912 Willis H. Carrier, Buffalo Forge Co., Buffalo, N.Y. Invented spray-type central station dehumidifier with Baudelot coils located below spray chamber. Filed Jan. 9, 1912. U.S. Patent #1,078,608, issued Nov. 18, 1913.

1912 Stuart W. Cramer, Charlotte, N.C. Installed central station spray-type humidifying system in Anderson Cotton Mill #2, Anderson, S.C.

1913 Dr. Leonard Hill and others, London Hospital Medical College. Reported tests on effects of atmosphere on test subjects in confined and crowded rooms. (70)

1913 J. M. Larson Introduced manually operated domestic refrigerating machine. (67)

1913 Vilter Manufacturing Company, Milwaukee. Supplied refrigerating system for air conditioning installation in Wisconsin Hotel, Milwaukee.

1913 Daniel M. Luehrs, American Blower Corporation, Detroit. Invented "Spraying Device and Means for Controlling Same," U.S. Patent #1,154,707, issued Sept. 28, 1915.

1913 Carrier Air Conditioning Company of America, New York. Installed comfort, central station, spray-type air conditioning system for dining room of Wisconsin Hotel, Milwaukee. Vilter Manufacturing Company, Milwaukee furnished refrigerating system.

RELATED DEVELOPMENTS AND THEORIES	HEATING AND VENTILATING FANS, HEATERS, HEAT PUMPS	REFRIGERATION (NOT INCLUDING COMMERCIAL EQUIPMENT AS DOMESTIC REFRIGERATORS, ETC.)	AIR CLEANING, HUMIDIFYING, PURIFYING. WASHERS, HUMIDIFYING UNITS OR HEADS, ODOR ABSORBERS, ETC.	AIR COOLING. FAN-ICE, COOLING-COIL AND FAN-COOLING-COIL UNITS AND SYSTEMS	AIR CONDITIONING. SPRAY TYPE CENTRAL STATION APPARATUS AND CHEMICAL DEHUMIDIFIERS
YEAR	YEAR	YEAR	YEAR	YEAR	YEAR
1914 *Fan Engineering*, edited by Willis H. Carrier. Published by Buffalo Forge Company, Buffalo. Publication in fifth edition, 1952.	1914 Albert A. Criqui, Buffalo Forge Co., Buffalo, N.Y. Filed first of five inventions, "Centrifugal Fan." Filed Sept. 3, 1914. U.S. Patent #1,161,926, issued Nov. 30, 1915.	1914 Frick Company, Waynesboro, Pa. Supplied refrigerating system for air conditioning installation, Beechnut Packing Co., Canajoharie, N.Y. 1914 Kroeschell Bros. Ice Machine Co., Chicago. Supplied refrigerating system for air conditioning installation, Sharpe & Dohme Co., Baltimore, Md. (See column 6, 1914.)			1914 Carrier Air Conditioning Company of America, New York. Installed year-round air conditioning system, Beechnut Packing Co., Canajoharie, N.Y. Frick Company, Waynesboro, Pa., supplied refrigerating system. 1914 Sharpe & Dohme Company, Baltimore installed year-round Carrier spray-type central station air conditioning system for capsule room. Kroeschell Bros. Ice Machine Co., supplied refrigerating system.
1915 Ellsworth Huntington, Professor of Yale University, wrote *Civilization and Climate*. Published by Yale University Press, New Haven, Conn. Revised 1924. 1915 Sweet's Catalogue Service, New York. For first time included index "Air Conditioning." Listed: American Blower Corporation, Carrier Air Conditioning Company			1915 Carrier Air Conditioning Company of America, New York (founded 1908) totally owned by Buffalo Forge Co., Buffalo, concentrated on air-washing and heating and ventilating systems.		1915 Carrier Air Conditioning Company of America, New York. By July 1915 had installed a total of approximately 500 central station air conditioning systems in a wide variety of industrial plants. 1915 Carrier Engineering Corporation, New York. Founded by Willis H. Carrier, J. Irvine Lyle, and five associates,

of America, Stuart W. Cramer, Spray Engineering Co., B. F. Sturtevant Co.

E. P. Heckel, L. L. Lewis, E. T. Lyle, E. T. Murphy and A. E. Stacey, Jr. As independent company concentrated on air conditioning systems, continuing to sell equipment built by Buffalo Forge Company, Buffalo, N.Y.

1916 Warren Webster & Company, Camden, N.J. Relinquished its air conditioning activities.

1916 Braemer Air Conditioning Corporation, Philadelphia. Founded in 1916 by Wm. G. R. Braemer, formerly with Warren Webster & Company, Camden, N.J.

1916 Carrier Engineering Corporation, New York. Purchased Braemer Air Conditioning Corporation. In 1919 changed its name to Atmospheric Conditioning Corporation. (Dissolved November 16, 1934.)

1916 American Blower Corporation, Detroit. Installed complete year-round central station spray-type air conditioning system in matchmaking industry.

RELATED DEVELOPMENTS AND THEORIES	HEATING AND VENTILATING FANS, HEATERS, HEAT PUMPS	REFRIGERATION (NOT INCLUDING COMMERCIAL EQUIPMENT AS DOMESTIC REFRIGERATORS, ETC.)	AIR CLEANING, HUMIDIFYING, PURIFYING, WASHERS, HUMIDIFYING UNITS OR HEADS, ODOR ABSORBERS, ETC.	AIR COOLING. FAN-ICE, COOLING-COIL AND FAN-COOLING-COIL UNITS AND SYSTEMS	AIR CONDITIONING. SPRAY TYPE CENTRAL STATION APPARATUS AND CHEMICAL DEHUMIDIFIERS
YEAR	YEAR	YEAR	YEAR	YEAR	YEAR
1917 Willis H. Carrier, Carrier Engineering Corporation, New York. Invented "Method of and Apparatus for Drying, Conditioning and Regulating Moisture Content of Hygroscopic Materials," known as "ejector system." Invention plays important part in many industrial and comfort air conditioning systems. Filed March 27, 1917. U.S. Patent #1,393,086, issued October 11, 1921.	1917 National Association of Fan Manufacturers. Founded 1917.				1917 Ilg Electric Ventilating Company, Chicago. Introduced unit spray-type apparatus with arrangement for cooling and dehumidifying air as well as humidifying and heating air. 1917 American Blower Corporation, Detroit. Installed complete year-round central station air conditioning system in Empire Theatre, Montgomery, Ala. —a 900-seat theater.
1918 John M. Frank, Ilg Electric Ventilating Company, Chicago. Invented "Unit for Heating and Ventilating Systems," disc-fan unit heater. Filed Jan. 9, 1918. U.S. Patent #1,295,151, issued Feb. 25, 1919.		1918 The Kelvinator Company (now Kelvinator Division, Nash-Kelvinator Corporation, Detroit). Sold first domestic refrigerating machines to U.S. markets. Engineered by E. J. Copeland. (67) 1918 Frigidaire Division, General Motors Corporation, Dayton, Ohio. Organized in 1918. Began manufacturing air-cooling units about 1928.	1918 Stuart W. Cramer (1868–1950) Charlotte, N.C. Retired in 1918. 1918 Parks-Cramer Company, Fitchburg, Mass., and Charlotte, N.C., formed G. M. Parks Co. (organized 1872) purchased Stuart W. Cramer interest. 1918 William B. Hodge (1872-?). Formerly with Stuart W. Cramer; joined Parks-Cramer Company.		1918 American Blower Corporation, Detroit. Installed central station spray-type air conditioning system for cheese curing and for lithographing plant.

1919 American Society of Heating and Ventilating Engineers established its Research Laboratory. (Originally located at U.S. Bureau of Mines, Pittsburgh, Pa.) Organizing committee: Wm. H. Driscoll, H. P. Gant, O. P. Hood, S. R. Lewis, Thornton Lewis, J. Irvine Lyle, and F. R. Still.

1919 The Niagara Blower Company, Chicago. Founded 1919. Extended activities to central station air conditioning systems and air-cooling units in 1926.

1919 York Heating & Ventilating Corporation, Philadelphia. Formed by combining York Heating and Ventilating Company (1910) and Lewis, Robinson and Gant (1912). Corporation in 1930 merged with others to form Carrier Corporation.

1919 (Circa) Wittenmeier Machine Company, Chicago. (See column 5, Balaban and Katz, 1919 and 1920.)

1919 Balaban and Katz. Installed fan-coil air-cooling system in Central Park Theater, Chicago. Wittenmeier Machine Co., supplied CO$_2$ refrigeration.

1919 Walter A. Patrick, Baltimore, Maryland. Invented "Silica Gel and making same." Filed Dec. 7, 1918. U.S. Patent #1,297,724 issued March 18, 1919.

1920-21 Arthur B. Modine, Modine Manufacturing Co., Racine, Wisc. Changed from manufacturing automobile radiators to building unit heaters. Employed many features of radiator in new product.

1920 Edward T. Curran, Detroit. Invented heating and cooling unit. (See column 5.)

1920 J. & E. Hall, Ltd., Dartford, Kent, England. Supplied refrigerating system for mine air conditioning of St. John del Rey Mining Co., Ltd., South America.

1920 Servel Inc., Evansville, Indiana. Founded 1920. (Entered air-cooling industry in 1934.)

1920 Edward T. Curran, Detroit. Invented "Means for Varying the Temperature of the Atmosphere." Automobile type radiator with disc-fan, cooled hot or cool medium in coil for temperature control. U.S. Patent #1,480,671, issued Jan. 25, 1924.

1920 Balaban & Katz. Installed fan-coil air-cooling system in Riviera Theatre, Chicago. CO$_2$ refrigeration supplied by Wittenmeier Machine Co., Chicago.

1920 J. & E. Hall, Ltd., Dartford, Kent, England. Installed central station air conditioning system and refrigeration to cool a mine for St. John del Rey Mining Co., Ltd., South America.

1920 Carrier Engineering Company, Ltd., London, England. Founded 1920. Stanley Groom, President.

RELATED DEVELOPMENTS AND THEORIES	HEATING AND VENTILATING FANS, HEATERS, HEAT PUMPS	REFRIGERATION (NOT INCLUDING COMMERCIAL EQUIPMENT AS DOMESTIC REFRIGERATORS, ETC.)	AIR CLEANING, HUMIDIFYING, PURIFYING. WASHERS, HUMIDIFYING UNITS OR HEADS, ODOR ABSORBERS, ETC.	AIR COOLING. FAN-ICE, COOLING-COIL AND FAN-COOLING-COIL UNITS AND SYSTEMS	AIR CONDITIONING. SPRAY TYPE CENTRAL STATION APPARATUS AND CHEMICAL DEHUMIDIFIERS
YEAR	YEAR	YEAR	YEAR	YEAR	YEAR
1922 American Society of Heating and Ventilating Engineers. Published first edition of the *Guide*.	1921 Harry S. Wheller, L. J. Wing Mfg. Co., Linden, N.J. Introduced disc-fan-type ceiling-suspended unit heater. Vento coil used in first model, changed to extended-surface coil in 1922. Revised unit's trade name, "Featherweight." 1922 York Heating and Ventilating Corporation, Philadelphia. Introduced a centrifugal-fan-type unit heater. Designated as "The Trail Blazer." Unit is in permanent exhibit, Franklin Institute, Philadelphia. 1922 Willis H. Carrier, Carrier Engineering Corporation, Newark, N.J. Invented an extended-surface heating coil. (See patents in appendix.) 1922 Lawrence C. Soule, Carrier	1921 Willis H. Carrier, Carrier Engineering Corporation, Newark, N.J. Invented Centrifugal Refrigerating Machine. (See list of patents in appendix.) 1922 Carrier Engineering Corporation, Newark, N.J. Demonstrated operation of centrifugal refrigerating machine at meeting of engineers and architects. 1922 The Carbondale Machine Company, Carbondale, Pa. Supplied refrigerating system for air conditioning system, Grauman's Metropolitan Theatre, Los Angeles. (See column 6, 1922.)	1921 Daniel M. Luehrs, American Blower Corporation, Detroit. Invented unit humidifier. 1922 A.S.H.V.E. *Guide*. Listed seven companies as supplying air washers: American Blower Corp., Bahnson Co., Bayley Mfg. Co., Buffalo Forge Co., Carrier Air Conditioning Co. of America, Hersh Bros. Co., and B. F. Sturtevant Co. 1922 Henderson and Haggard patented "deodorizing of offensive gaseous emanations from organic matter." U.S. Patent #1,410,-249. (22)	1922 Jewell B. Williams. Wrote paper on air cooling by refrigeration (bunker coils) for industry and comfort. (72)	1921 J. O. Ross Engineering Corporation, Division of Ross Industries Corporation, New York. Founded 1921. 1921 American Blower Corporation, Detroit. Installed central station air conditioning system for sausage production. 1922 J. O. Ross Engineering Corporation, New York. Installed year-round air conditioning system for candy plants. 1922 A.S.H.V.E. *Guide*. Listed six companies as suppliers of air conditioning apparatus: American Blower Corp., Atmospheric Conditioning Corp., Carrier Air Conditioning Co. of America, Carrier Engineering Corp., Drying Systems, Inc., and W. L. Fleisher & Co.

1922 Grauman's Metropolitan Theater, Los Angeles, Calif. Installed central station spray-type air conditioning system with down-draft by-pass air distribution. The Carbondale Machine Company, Carbondale, Pa., supplied CO_2 refrigerating system. Theater opened January 26, 1923.

Engineering Corporation, Newark, N.J. Invented heaters with extended-surface coil, later named "Aerofin." Filed Nov. 8, 1922. U.S. Patent #1,597,733, issued August 31, 1926.

1922 T. B. Morley Wrote on theory of heat pump—had idea for present-day developments. (11)

1923 A.S.H.V.E. Research Laboratory, U.S. Bureau of Mines, Pittsburgh. Established "Effective Temperature Lines." Research performed by F. C. Houghton, C. P. Yaglou, Dr. W. J. McConnell under Director of Laboratory, Dean F. Paul Anderson.

1922 Fred Wittenmeier, Wittenmeier Machine Co., Chicago. Installed air-cooling (coil and spray combined) apparatus in theaters and public buildings. (71) Spray on coil to increase heat transfer—not for dew-point control.

1922 Brunswick - Kroeschell Company, formed in 1922 by merger of Kroeschell Bros. Ice Machine Co. and Brunswick Machine Co. In 1930 company became part of Carrier Corporation.

1923 Aerofin Corporation (Newark, N.J.). Founded 1923. Originally called Soule Heater Corporation.

1923 Henry Baetz, St. Louis. Invented "Air Heater," multi-vane fan, downward discharge, two directional outlets. Filed May 1, 1923. U.S. Patent #1,517,487, issued Dec. 2, 1924. (Also see invention above, issued 1898.)

1923 Arthur B. Modine, Racine, Wisc. Invented "Heating Unit," disc-fan type. Filed April 7, 1923. U.S. Patent #1,666,907, issued April 24, 1928.

1923 American Blower Corporation, Detroit. Installed central station spray-type air conditioning systems in textile mills.

1923 Stephen F. Whitman & Sons, Philadelphia. Signed contract to install three Carrier Centrifugal Refrigerating Machines to chill water for air conditioning system installed by Carrier Engineering Corporation.

1923 W. F. Schrafft & Sons Co., Boston. Signed contract for one Carrier Centrifugal Refrigerating machine to chill water for air con-

1923 Carrier Engineering Corporation, Newark, N.J. Supplied three centrifugal refrigerating machines for air conditioning system in Stephen F. Whitman & Sons candy plant, Philadelphia. (In 1952 company operates 4 machines of this type, totaling 1282 tons.)

1923 Carrier Engineering Corporation, Newark, N.J. Supplied one refrigerating machine to chill water for air conditioning system installed at W. F. Schrafft & Sons Co., Boston.

RELATED DEVELOPMENTS AND THEORIES	HEATING AND VENTILATING FANS, HEATERS, HEAT PUMPS	REFRIGERATION (NOT INCLUDING COMMERCIAL EQUIPMENT AS DOMESTIC REFRIGERATORS, ETC.)	AIR CLEANING, HUMIDIFYING, PURIFYING. WASHERS, HUMIDIFYING UNITS OR HEADS, ODOR ABSORBERS, ETC.	AIR COOLING. FAN-ICE, COOLING-COIL AND FAN-COOLING-COIL UNITS AND SYSTEMS	AIR CONDITIONING. SPRAY TYPE CENTRAL STATION APPARATUS AND CHEMICAL DEHUMIDIFIERS
YEAR	YEAR	YEAR	YEAR	YEAR	YEAR
1924 L. Logan Lewis, Carrier Engineering Corporation, Newark, N.J. Invented "Apparatus for Cooling and Ventilating." Later known as the "By-pass" system for air distribution. (See Auditorium Conditioning Corporation, 1927 below.) Invention filed Dec. 29, 1924. U.S. Patent #1,583,060, issued May 4, 1926.	1923 John M. Frank, Ilg Electric Ventilating Company, Chicago. Invented a draw-through disc-fan-type unit heater. U.S. Patent #1,626,400, issued April 26, 1927. 1924 E. H. Seelert, McQuay, Inc., Minneapolis, Minnesota. Designed a unit heater. (Company entered air-cooling market in 1936.)	1923 Willis H. Carrier, Carrier Engineering Corporation, Newark, N.J. Invented a new refrigerant (Dielene) suitable for centrifugal compression. (See list of patents.)			...ditioning system installed by Carrier Engineering Corporation. 1924 Pantheon Theater, Chicago. Installed central station spray-type air conditioning system. Walter L. Fleisher spray apparatus, B. F. Sturtevant Co. fans, Brunswick-Kroeschell Co. refrigeration. Theater: 20,000 seats, 60,000 cfm. 1924 Thomas Chester, consulting engineer (see 1903 column 5). Designed year-round spray-type air conditioning system for Ohio Match Co., Wadsworth, Ohio. Cooling capacity totaled 900 tons.

1925 Willis H. Carrier, Carrier Engineering Corporation, Newark. Designed centrifugal refrigerating machine to make ice for Madison Square Garden, New York, skating floor.
1925 Brown - Boveri. Designed centrifugal refrigerating machine employing ammonia as refrigerant. (73)

1925 Dr. E. Vernon Hill, Ventilating Division, City Health Department, Chicago, Ill. Proposed "Don'ts" on Theater Ventilation. Mentioned "coolers" added to fan, heating, ventilating, washing systems. Cites Riviera Theatre, Chicago, as first to experiment with refrigerating coils in such an installation. (75)

1925 Drying agent, Silica Gel, first applied to air conditioning systems. (Silica Gel was discovered by Walter A. Patrick in World War I research at Johns Hopkins University, Baltimore.) (74)
1925 The United States Air Conditioning Corporation, Minneapolis, Minn. Founded 1925.
1925 The United States Air Conditioning Corporation, Minneapolis, Minn. Installed evaporative-cooling central station spray-type air conditioning system in theaters. Between 1925 and 1930 the company's subsidiaries installed some 8,000 such systems.

1926 A. R. Stevenson, General Electric Company, Schenectady, N.Y. Suggested that refrigerating machines be employed as heat pumps. (76)

1926 Willis H. Carrier, Carrier Engineering Corporation, Newark, N.J. Introduced new refrigerant, Carrene 1. (See list of patents.)
1926 Harry E. Thompson, Detroit. Invented a thermostatic expansion valve. Assigned to Universal Cooler, Div., Tecumseh Products Co., Marion, Ohio. Filed Feb. 15, 1926. U.S. Patent #1,634,633, issued July 5, 1927.

RELATED DEVELOPMENTS AND THEORIES	HEATING AND VENTILATING FANS, HEATERS, HEAT PUMPS	REFRIGERATION (NOT INCLUDING COMMERCIAL EQUIPMENT AS DOMESTIC REFRIGERATORS, ETC.)	AIR CLEANING, HUMIDIFYING, PURIFYING. WASHERS, HUMIDIFYING UNITS OR HEADS, ODOR ABSORBERS, ETC.	AIR COOLING. FAN-ICE, COOLING-COIL AND FAN-COOLING-COIL UNITS AND SYSTEMS	AIR CONDITIONING. SPRAY TYPE CENTRAL STATION APPARATUS AND CHEMICAL DEHUMIDIFIERS
YEAR	YEAR	YEAR	YEAR	YEAR	YEAR
1927 Auditorium Conditioning Corporation, New York. Incorporated in 1927. (Dissolved 1946.)	1927 Young Radiator Company, Racine, Wisc. Founded 1927, Fred M. Young, president.		1927 Howard C. Murphy, American Air Filter Company, Louisville, Ky. Reported on advances in design and application of oil-coated filters. (78) 1927 Wm. G. R. Braemer, Universal Humidifier Corporation, Philadelphia. Designed electrical humidifier for individual rooms.		1927 Willis H. Carrier, Carrier Engineering Corporation, Newark, N.J. Developed home air conditioning unit for winter heating, humidifying, and cleaning air for residences.
				1928 H. P. Gant, York Heating and Ventilating Corporation, Philadelphia. Invented coil-fan unit, a "cold diffuser." Filed Aug. 8, 1928, U.S. Patent #1,860,357, issued May 31, 1932. 1928 Ilg Electric Ventilating Company, Chicago. Introduced "Ilg Airator," an evaporative-cooling unit believed to be forerunner of such units used extensively in Southwestern United States.	1928 Carlyle M. Ashley, Carrier Engineering Corporation, Newark, N.J. Invented spray-type vertical unit air conditioner, the "Centriector." Filed April 9, 1928. U.S. Patent #1,883,456, issued Oct. 18, 1932. 1928 Carrier-Lyle Corporation, a subsidiary of Carrier Engineering Corporation, Newark, N.J. Incorporated Oct. 23, 1928. Handled winter air conditioning in residences. (Dissolved Jan. 4, 1935.)

1928 Claude A. Bulkeley, Niagara Blower Company, Buffalo, N.Y. Invented a vertical spray-type air conditioning unit. Filed Dec. 14, 1928. U.S. Patent #1,770,765, issued July 15, 1930. First of many inventions by him to advance the art of air conditioning.

1929 Baltimore & Ohio Railroad, Baltimore, Md. Installed trial equipment for spray-type air conditioning tests on a car in yards. System designed by Willis H. Carrier, Carrier Engineering Corporation, Newark, and tested under his supervision. (System later adopted for railroad cars was fan-coil cooling units.)

1929 General Electric Company, Schenectady, N.Y. Built self-contained water-cooled room cooler; sulphur dioxide. Tested unit in home of Willis H. Carrier, Essex Falls, N.J., summer, 1929.

1930 Niagara Blower Corporation, Buffalo, N.Y. Installed year-round, spray-type, air conditioning system. Refrigeration, ammonia system, to chill

1930 Room coolers subject of development by many companies. First types were room coolers with remote refrigerating machines.

1930 Development of "Freon-12" announced. Dr. Thomas Midgley, Jr., and Dr. Albert L. Henne with General Motors and E. I. du Pont de

1929 Alfred E. Stacey, Jr., Carrier Engineering Corporation, Newark, N.J. Invented method for distributing air for railroad car cooling. Filed Nov. 29, 1929. U.S. Patent #1,982,125, issued Nov. 27, 1934.

1929 Willis H. Carrier, Carrier Engineering Corporation, Newark. Applied ejector nozzles in room discharge for induced secondary air flow and thus began development that led to Conduit Weathermaster System. Installed test unit in company's directors' room, 850 Frelinghuysen Ave., Newark.

RELATED DEVELOPMENTS AND THEORIES	HEATING AND VENTILATING FANS, HEATERS, HEAT PUMPS	REFRIGERATION (NOT INCLUDING COMMERCIAL EQUIPMENT AS DOMESTIC REFRIGERATORS, ETC.)	AIR CLEANING, HUMIDIFYING, PURIFYING. WASHERS, HUMIDIFYING UNITS OR HEADS, ODOR ABSORBERS, ETC.	AIR COOLING. FAN-ICE, COOLING-COIL AND FAN-COOLING-COIL UNITS AND SYSTEMS	AIR CONDITIONING. SPRAY TYPE CENTRAL STATION APPARATUS AND CHEMICAL DEHUMIDIFIERS
YEAR	YEAR	YEAR	YEAR	YEAR	YEAR
		Nemours research group produced the new and revolutionary refrigerant. Manufactured by Kinetic Chemicals, Inc. (now Kinetic Chemicals Division of E. I. du Pont de Nemours), Wilmington, Delaware. (79)		1930 Baltimore & Ohio Railroad, Baltimore, Md. Installed an air-cooling system on Martha Washington diner. System designed by Carrier Engineering Corporation. Refrigerating machine supplied by Brunswick-Kroeschell Company.	water, supplied by Baker Refrigeration Corporation, South Windham, Maine.
		1930 Baker Refrigeration Corporation, South Windham, Maine. Supplied refrigerating system for air conditioning apparatus built by Niagara Blower Corporation, Buffalo.		1930 Thornton Lewis, York Heating and Ventilating Corporation, Philadelphia. Invented disc-fan-type cold diffuser. Filed July 28, 1930. U.S. Patent ※1,917,043, issued July 4, 1933.	1930 Carrier Corporation, Newark, N.J. Formed by merger of Carrier Engineering Corporation, Brunswick-Kroeschell Company, and York Heating and Ventilating Corporation.
				1930–1931 Lehman Bros., New York. Purchased room coolers with remote refrigerating machines. York Heating and Ventilating Corporation supplied room units. Frigidaire Division, General Motors supplied refrigerating machines. Refrigerant changed from sulphur dioxide to "Freon-12" during installation.	

1931 Axial-flow fans, originating in aeronautical research, adapted to ventilating industry by English engineers. (80)

1931 Albert R. Thomas, Servel, Inc., Evansville, Ind. Invented "Refrigerating System." Filed Sept. 18, 1931. U.S. Patent #1,964,391, issued June 26, 1934.

1932 *Heat Transmission* by Prof. William H. McAdams. Published by McGraw-Hill Publishing Company, New York. Written under auspices of National Research Council Committee on Heat Transfer, Willis H. Carrier, Chairman. (Second edition published 1942.)

1932 Carlyle M. Ashley, Carrier Corporation, Newark, N.J. Invented "Air Conditioning System." Employed steam for heating and cooling in steam-ejector system. Filed August 13, 1932. U.S. Patent #2,010,001, issued August 6, 1935.

1932 Charles R. Neeson (later with Air-temp Division, Chrysler Corporation, Dayton, Ohio). Began research on design of radial refrigerating compressors. (81)

1932 Henry Galson and Charles R. Neeson, Baldwin-Southworth Corporation, Philadelphia. Invented self-contained air-cooled "Freon-12" room coolers. Awarded John Scott medal for invention.

1932 Baker Refrigeration Corporation, South Windham. Introduced its "Cold Stream Cooling Units."

1933 Ingersoll-Rand Company, Phillipsburg, N.J. (Established 1871.) Entered refrigerating industry in 1933. Built and installed water-vapor centrifugal refrigerating machines to chill water for air conditioning systems. Also built and installed steam-ejector refrigerating machines. (82)

1933 John M. Frank, Ilg Electric Ventilating Company, Chicago. Invented "Ilg Spot Cooler." Sold several thousand in 1930's. Filed Sept. 30, 1933. U.S. Patent #2,084,392, issued June 22, 1937. (Water cooled condensers.)

1933 Ingersoll-Rand Company, Philadelphia. Installed water-vapor centrifugal refrigerating machine to chill water for air conditioning G. C. Murphy Store, Washington, D.C.

RELATED DEVELOPMENTS AND THEORIES	HEATING AND VENTILATING FANS, HEATERS, HEAT PUMPS	REFRIGERATION (NOT INCLUDING COMMERCIAL EQUIPMENT AS DOMESTIC REFRIGERATORS, ETC.)	AIR CLEANING, HUMIDIFYING, PURIFYING. WASHERS, HUMIDIFYING UNITS OR HEADS, ODOR ABSORBERS, ETC.	AIR COOLING. FAN-ICE, COOLING-COIL AND FAN-COOLING-COIL UNITS AND SYSTEMS	AIR CONDITIONING. SPRAY TYPE CENTRAL STATION APPARATUS AND CHEMICAL DEHUMIDIFIERS
YEAR	YEAR	YEAR	YEAR	YEAR	YEAR
1935 American Blower Corporation and Canadian Sirocco Company, Ltd. Published *Air Conditioning and Engineering.* (359 pages.)	1935 General Electric Company, Bloomfield, N.J. Installed Heat Pump in Atlantic City Electric Company office building in Salem, N.J. (76)	1933 Carlyle M. Ashley, Carrier Corporation, Newark, N.J. Invented evaporative condenser, disc-fan-type. Filed Nov. 15, 1933. U.S. Patent #2,059,839, issued Nov. 3, 1936. 1933 Brown-Boveri. (Switzerland.) Began building centrifugal refrigerating machines with Carrene 1 as refrigerant. (73) 1933 Willis H. Carrier, Carrier Corporation, Newark, N.J. Introduced Carrene 2 as refrigerant. (See list of patents.) 1935 General Electric Company, Bloomfield, N.J. (See Heat Pump, column 2, 1935.)	1935 Gaylord W. Penney, Westinghouse Electric & Manufacturing Company, East Pittsburgh, Pa. Invented "Electrical Precipitation for Atmospheric Dust." Filed Oct. 15, 1935. U.S. Patent #2,129,783, issued Sept. 13, 1938. (83)	1934 Servel, Inc., Evansville, Ind. Installed floor-type comfort-cooling units. 1935 General Electric Company, Bloomfield, N.J. (See Heat Pump, column 2, 1935.)	

1936 Eugene F. Du Bois, Physiologist, New York. Carried forward research on human body heat losses and temperature regulation in conditioned air. (84)

1937 Willis H. Carrier, Carrier Corporation, Syracuse, N.Y. Designed "Conduit Weathermaster System," High pressure supply air and secondary induced air flow.

1938 Prof. C. E. A. Winslow, L. P. Herrington and A. P. Gagge, Yale University, New Haven, Conn., reported on research on reactions of clothed human body to varying atmospheric humidities. (86)

1938 C. O. Mackey, Cornell University, Ithaca, N.Y. Studied air jet in improving performance of air distributing outlets. (87)

1939 Walter Jones, Carrier Corporation, Syracuse, N.Y. Invented system for refrigerating and heating (reverse cycle or heat pump) with Carrier

1936 Walter Jones, Carrier Corporation, Newark, N.J. Applied centrifugal refrigerating machine to condense ammonia. Filed Dec. 8, 1936. U.S. Patent #2,145,692, issued Jan. 31, 1939. Patent applied in P. Ballantine & Sons, Newark, N.J., brewery in 1936.

1938 The Trane Company, La Crosse, Wisconsin. (Founded 1886.) Began building centrifugal refrigerating machines, "Freon-113" as refrigerant. (73)

1939 York Corporation, York, Pa. (Founded 1874.) Began building centrifugal refrigerating machines, Carrene 2 as refrigerant. (73)

1936 George S. Dauphinee, W. B. Connor Engineering Corporation, New York. Invented "Odor Removing Apparatus"—Activated Carbon. Filed Sept. 3, 1936. U.S. Patent #2,214,737, issued Sept. 17, 1940. Was issued additional patents on odor-absorbing apparatus in 1942.

1937 Parks-Cramer Company, Fitchburg, Mass. Introduced self-cleaning atomizing spray in unit humidifier.

1939 American Air Filter Company, Louisville, Ky. (Founded 1927.) Began building electrostatic precipitators. (88)

1939 Woodward Iron Company, Woodward, Ala. Installed blast-furnace air conditioning system, and opened up market for such applications. In-

RELATED DEVELOPMENTS AND THEORIES	HEATING AND VENTILATING FANS, HEATERS, HEAT PUMPS	REFRIGERATION (NOT INCLUDING COMMERCIAL EQUIPMENT AS DOMESTIC REFRIGERATORS, ETC.)	AIR CLEANING, HUMIDIFYING, PURIFYING. WASHERS, HUMIDIFYING UNITS OR HEADS, ODOR ABSORBERS, ETC.	AIR COOLING. FAN-ICE, COOLING-COIL AND FAN-COOLING-COIL UNITS AND SYSTEMS	AIR CONDITIONING. SPRAY TYPE CENTRAL STATION APPARATUS AND CHEMICAL DEHUMIDIFIERS
YEAR	YEAR	YEAR	YEAR	YEAR	YEAR
	centrifugal refrigerating machine. Filed Jan. 18, 1939. U.S. Patent #2,219,815, issued Oct. 29, 1941.	1939 Walter Jones, Carrier Corporation, Syracuse. Issued U.S. Patent on Reverse Cycle. (See column 2, 1939.)			active since 1912 or 1913. Carrier Corporation supplied central spray-type air conditioning apparatus and centrifugal refrigerating system for this installation.
					1939 Cia Textilera Ari Guanabo, S.A., Havana, Cuba. Installed mechanical refrigeration with air conditioning apparatus to control both temperature and humidity in a cotton mill; opened market for such systems. Carrier Corporation supplied central station air conditioning system and centrifugal refrigerating system to cool spray water.
1940 *Modern Air Conditioning, Heating and Ventilating* by Willis H. Carrier, Realto E. Cherne and Walter A. Grant. Published by Pitman Publishing Corporation, N.Y. Copyright 1940. Second edition, 1950. 1940 L. E. Seeley, University of New		1940 Walter Jones, Carrier Corporation, Syracuse, N.Y. Applied extruded tubes to refrigerating condensers and evaporators. Filed June 20, 1940. U.S. Patent #2,314,402, issued March 23, 1943. Known as "Lo-Fin" tubes.			

Hampshire, Durham, N.H. Reported on research on temperature changes and moisture content in respired air. (89) (84)

1942 G. L. Tuve, G. B. Priester, and D. K. Wright, Jr., A.S.H.V.E. Research Laboratory, Cleveland, Ohio. Reported on jet action in air distribution research. (90)

1944 G. L. Tuve and G. B. Priester, A.S.H.V.E. Research Laboratory, Cleveland. Reported on control of air streams in large spaces. (91)

1946 Alexis A. Berestneff, Carrier Corporation, Syracuse. Invented means to apply absorption principle of refrigeration to large-capacity machines, water as refrigerant. Filed in 1946. U.S. Patent #2,461,-513, issued February 15, 1949. (92)

1946 Winston H. Reed and William A. Pennington, Carrier

1942 National Advisory Committee on Aeronautics. Installed air conditioning system in wind tunnel at Cleveland, Ohio. Carrier Corporation supplied refrigerating system to reduce air to −67 F. Nominal rating total capacity of installation equals approximately 20,000 tons.

AIR CONDITIONING. SPRAY TYPE CENTRAL STATION APPARATUS AND CHEMICAL DEHUMIDIFIERS	AIR COOLING. FAN-ICE, COOLING-COIL AND FAN-COOLING-COIL UNIT SYSTEMS	AIR CLEANING, HUMIDIFYING, PURIFYING. WASHERS, HUMIDIFYING UNITS OR HEADS, ODOR ABSORBERS, ETC.	REFRIGERATION (NOT INCLUDING COMMERCIAL EQUIPMENT AS DOMESTIC REFRIGERATORS, ETC.)	HEATING AND VENTILATING FANS, HEATERS, HEAT PUMPS	RELATED DEVELOPMENTS AND THEORIES
YEAR	YEAR	YEAR	YEAR	YEAR	YEAR
1948 Cross Cotton Mills, Marion, N.C. Installed Carrier Corporation central station year-round air conditioning system and centrifugal refrigerating machine for control of temperature and humidity in cotton mill. 1952 Total manufacturers of central station spray-type air conditioning apparatus ..21 Total manufacturers of chemical dehumidifiers 9 (94)	1952 Total manufacturers of air-cooling apparatus: Self-contained type of units ....38 Units with remote refrigerating machines 68 (94)	1952 Total manufacturers of air-cleaning, purifying, dehumidifying, and humidifying apparatus: Filters (dust) ....17 Electric precipitators ....6 Odor Absorbers ....5 Humidifying Heads 11 Air Washers 21 (94)	Corporation, Syracuse, N.Y. Invented "Process for Producing Increased Refrigeration — Carrene-7." Filed May 10, 1946. U.S. Patent #2,479,259, issued Aug. 16, 1949. Refrigerant manufactured in 1950. (93) 1952 Total refrigerating machine manufacturers: Condensing Units, Air-Cooled ....33 Condensing Units, Water-Cooled .....34 Hermetic-Sealed Condensing Units....5 Industrial Refrigerating Systems 27 (94)	1952 Total fan manufacturers (not including disc fans) ..48 (94) Total unit heater manufacturers ....19 (95) Total manufacturers of heat pumps 16 (94)	1952 Total companies in air conditioning and refrigerating industry in United States (not including domestic and commercial refrigerating units) ..211 (94)

REFERENCES FOR CHRONOLOGICAL CHART

(1) *The Mechanical Investigations of Leonardo da Vinci* by Ivor B. Hart. Published by Chapman & Hall, Ltd., London, 1925. (Models of instruments in Collections of the Fine Arts, International Business Machines Corporation.)
(2) *Encyclopedia Britannica, 1946.*
(3) *Ventilation and Heating* by J. S. Billings. Published by Engineering Record, New York, 1893.
(4) Hyanson *Dictionary of Universal Biography.*
(5) *History and Art of Warming and Ventilating* Vol. II by Walter Berman. Published by George Bell, London, 1845.
(6) *A Practical Treatise on Heating* by Thomas Box. Published by E. F. N. Spon, London, 1880.
(7) *Theory and Practice of Filtration* by George D. Dickey and Charles L. Bryden. Modern Library of Chemical Engineering. Published by Reinhold Publishing Corporation, New York, 1946.
(8) "Mechanical Refrigeration. Its American Birthright" by W. R. Woolrich, *Refrigerating Engineering,* March 1947.
(9) "The Development of Refrigeration in the United States" by J. F. Nickerson. *Ice and Refrigeration,* October 1915.
(10) "Methods of Testing Blowing Fans" by R. C. Carpenter. A.S.H.V.E., Trans., 1900.
(11) *The Heat Pump* by J. B. Pinkerton. Published by Princes Press, Ltd., Westminster S.W.I. London, 1949.
(12) "Gorrie—Pioneer Ice Maker" (staff article), *Ice and Refrigeration,* July 1950.
(13) Irish Academy Transactions Vol. 17, p. 275, 1837. (Cited in "Rational Psychrometric Formulae" by Willis H. Carrier. A.S.M.E., Trans., 1911.)
(14) *The Practical Methods of Ventilating Buildings* by Luther V. Bell. Address delivered before Massachusetts Medical Society, May 31, 1848. Published by Dickenson Printing, printed by Damrell & Moore, Boston.

(15) *Practical Treatise on Ventilation* by Morill Wyman. Published by James Munroe and Company, Boston, and Chapman Brothers, London, 1846.

(16) "Early Comfort Cooling Plants" by G. Richard Ohmes and Arthur K. Ohmes. *Heating, Piping and Air Conditioning,* June 1936.

(17) Phil. Transactions, Royal Society, 1851, p. 141. (Cited in "Rational Psychrometric Formulae" by Willis H. Carrier. A.S.M.E., Trans., 1911.)

(18) "The Centrifugal Fan" by Frank L. Busey, A.S.H.V.E., Trans., 1915.

(19) Kents *Mechanical Engineers Pocket-Book,* copyright 1895 (1900 printing).

(20) "Psychrometric Tables for Obtaining the Vapor Pressure, Relative Humidity, and Temperature of the Dew Point" by C. F. Marvin, U. S. Weather Bureau, 1940.

(21) "Description des Machine et Procédés pour lesguels des Brevets d'Invention," Tome 52. (Reference from Wallerstein Laboratories, New York City.)

(22) *Odors—Physiology and Control* by Carey P. McCord and William N. Witheridge. Published by McGraw-Hill Book Company, New York, 1949.

(23) "Some Recollections of the Refrigerating Industry" by Eugene T. Skinkle. *Ice and Refrigeration,* Nov. 1916.

(24) *Air Conditioning in Textile Mills,* edited by Albert W. Thompson, published by Parks-Cramer Co., 1925.

(25) "Remarks of Mr. J. G. Garland." New England Cotton Manufacturers' Association. Proceedings of the Semi-Annual Meeting, Boston, Oct. 29, 1879.

(26) Journal of the Patent Office Society Vol. XXXII, No. 1, Jan. 1950.

(27) "History of the Heat Regulator" by J. W. Pauling. *Heating and Ventilating,* June 1929.

(28) Advertisement, The Carbondale Machine Company. *Ice and Refrigeration,* Nov. 1915.

(29) *Air Conditioning and Engineering,* published by American Blower Corporation, 1935.

(30) "The Sturtevant System of Heating, Ventilating and Moistening" by Eugene N. Foss. New England Cotton Manufacturers' Association. Proceedings of the Twenty-fourth Annual Meeting, Boston, April 24, 1889.

(31) Annual Report, Ch. Signal Officer, 1886. (Cited in "Rational Psychrometric Formulae" by Willis H. Carrier. A.S.M.E., Trans., 1911.)

(32) Report of Air Conditioning and Refrigerating Machinery Association given at meeting of Refrigerating Machinery Association, May 1935. From copy lent by Mr. Wm. B. Henderson, Executive Vice President, Washington, D.C.

(33) *American Fabrics* by Bendure Pfeiffer, published by Macmillan Company, 1946.

(34) "25 Years of Fan History" by Chas. E. Lawrence. *Heating and Ventilating,* June 1929.

(35) *Mechanical Ventilation and Heating by a Forced Circulation of Warm Air* by Walter B. Snow. Lecture delivered at Sibley College, Cornell University, Nov. 17, 1899. Published by B. F. Sturtevant Co., Boston, Mass. Fourth Edition, 1907.

(36) Catalogue "Linde Canadian Refrigeration Company, Ltd." Montreal, about 1905 or 1906 describes Linde air cooler with brine sprayed over cooling coils. Brine spray prevents ice forming on coil from condensate.

(37) "Development of Pipe Line Refrigeration" by R. H. Tait. *Ice and Refrigeration,* Nov. 1916.

(38) "Refrigeration Twenty-five Years Ago" by John E. Starr. *Ice and Refrigeration,* Nov. 1916.

(39) "A Large Refrigerating Machine," *Scientific American,* April 16, 1892.

(40) "Application of Heat-unit System in Calculations" by Alfred R. Wolff. Franklin Institute, 1894. A.S.M.E., Trans., 1909.

(41) "Cooling of Air in Summer" by Charles H. Herter. *Southern Engineer,* October and November 1910.

(42) "Cooling an Auditorium by the Use of Ice" by John J. Harris. A.S.H.V.E., Trans., 1903.

(43) *The Whitin Machine Works Since 1831* by Thomas R. Navin, published by Harvard University Press, Cambridge, Mass., 1950.

(44) *Business Week,* Dec. 24, 1949.

(45) "The Dictionary of Electrical Words, Terms and Phrases" by E. J. Houston, 1897 (National Bureau of Standards, Washington, D.C.) used "English Heat Unit" in text, in appendix written for 1897 edition used "British Thermal Unit."

(46) "Development of Carbon Dioxide Refrigerating Machines" by F. Wittenmeier. *Ice and Refrigeration,* Nov. 1916.

(47) Kroeschell Brothers Ice Machine Company. Catalogue (not dated) latest testimonial letter in catalogue dated 1911.

(48) "New Epoch Ushered in by Unit Heaters" by Henry Baetz. *Heating and Ventilating,* June 1929.

(49) "Air Conditioning the New York Stock Exchange" by Donald A. Kepler. *Heating, Piping and Air Conditioning,* April 1947.

(50) "Air Conditioning Apparatus" by Willis H. Carrier and Frank L. Busey. A.S.M.E., Trans., 1911.

(51) "Industrial Research Laboratories of the United States." Published by the National Research Council, National Academy of Science, Washington, D.C. Eighth Edition, 1946.

(52) "Refrigeration in France" by L. Marchis. *Ice and Refrigeration*, Oct. 1915.

(53) "Our 25th Anniversary, 1904–1929" by A. L. Armagnac, *Heating and Ventilating*, June 1929.

(54) "Recollections of the Ice Machine Industry" by Thomas Shipley (Editors' introduction to article). *Ice and Refrigeration*, Nov. 1916.

(55) "Progress in Air Conditioning in the Last Quarter Century" by Willis H. Carrier. A.S.H.V.E., Trans., 1936.

(56) "Refrigeration for Blast Furnace." Editorial. *Ice and Refrigeration*, May 1905.

(57) "How Air Conditioning Has Developed in Fifty Years" by Walter L. Fleisher. *Heating, Piping and Air Conditioning*, Jan. 1950.

(58) "A Combination Ventilating, Heating and Cooling Plant in a Bank Building" by A. M. Feldman. A.S.H.V.E., Trans., 1909.

(59) *Cottrell; Samaritan of Science* by Frank Cameron. Published by Doubleday & Company, Garden City, New York, 1952.

(60) "Recent Developments in Air Conditioning" by Stuart W. Cramer, Charlotte, N.C. Paper read before convention of American Cotton Manufacturers Association, May 16–17, 1906.

(61) "Old Hands at Air Conditioning" by E. C. Thrasher, *Factory Management and Maintenance*, Feb. 1938.

(62) *Radiant Heating* by T. Napier Adlam. Published by The Industrial Press, New York, 1947.

(63) "Notes of a Pioneer in Air Conditioning" by Walter L. Fleisher, *Heating and Ventilating*, June 1929.

(64) "Commercial Uses of Refrigerating Machinery" by S. S. Vandervaart, *Ice and Refrigeration*, Nov. 1916.

(65) "Cast Iron Hot-Blast Heaters—New Methods in Testing and New Formula" by L. C. Soule. A.S.H.V.E., Trans., 1910.

(66) "Heat Transmission with Pipe Coils and Cast-iron Heaters under Fan Blast Conditions" by L. C. Soule. A.S.H.V.E., Trans., 1913.

(67) *Modern Electric and Gas Refrigeration* by Andrew D. Althouse and Carl H. Turnquist. Published by The Goodheart-Wilcox Company, Chicago, 1950.

(68) "Rational Psychrometric Formulae" by Willis H. Carrier. A.S.M.E., Trans., 1911.

(69) "The Aerodynamic Development of Axial Flow Fans" by T. H. Troller, A.S.H.V.E., Trans., 1944.

(70) "The Influence of the Atmosphere on our Health and Comfort in Confined and Crowded Places." Published by Smithsonian Institute, Washington, D.C., 1913.

(71) "Cooling of Theatres and Public Buildings" by Fred Wittenneyer. *Ice and Refrigeration*, July 1922.

(72) "Air Cooling by Refrigeration" by Jewell B. Williams. *Refrigerating World*, Dec. 1922.

(73) "A History of the Centrifugal Refrigeration Machine" by Walter A. Grant. *Refrigerating Engineering*, Feb. 1942.

(74) "Silica Gel as a Drying Agent" by R. L. Hockley. *Refrigerating Engineering*, April 1940.

(75) "Don'ts on Theatre Ventilation" by E. Vernon Hill. *Heating and Ventilating Magazine*, March 1925.

(76) "An All Electric Heating, Cooling and Air Conditioning System" by Philip Spon and D. W. McLenegan. A.S.H.V.E., Trans., 1935.

(77) "Silica Gel Developing into One of Chemistry's Great Contributions to Industry" by Richard Woods Edmonds, *Manufacturers Record*, May 16, 1928.

(78) "Design and Application of Oil Coated Filters" by H. C. Murphy. A.S.H.V.E., Trans., 1927.

(79) "Mark Air Conditioning's 50th Birthday This Year," *Heating, Piping and Air Conditioning*, Feb. 1952.

(80) "The Axial Flow Fan and Its Place in Ventilation" by W. R. Heath and A. E. Crique. A.S.H.V.E., Trans., 1944.

(81) "Variable Capacity Radial Compressors" by C. R. Neeson. *Refrigerating Engineering*, Feb. and March 1942.

(82) "Centrifugal Water-Vapor Refrigeration" by Paul Bancel, American Institute of Chemical Engineers, Trans., 1930.

(83) "A New Electrostatic Precipitator" by G. W. Penney, *Electrical Engineering*, Jan. 1937.

(84) "The Mechanism of Heat Losses and Temperature Regulation" by Eugene DuBois. Trans., Association of American Physicians, 1936.

(85) "American Developments in Air Conditioning and Refrigeration During the Last Decade" by Willis H. Carrier. The Institution of Mechanical Engineers (London) Proceedings, 1947.

(86) "The Reactions of the Clothed Human Body to Variations in Atmospheric Humidity" by C. E. A. Winslow, L. P. Herrington, and A. P. Gagge. *American Journal of Physiology*, Dec. 1938.

(87) "The Rationale of Air Distribution and Grille Performance" by C. O. Mackey. *Refrigerating Engineering*, June 1938.

(88) American Air Filters Company, Louisville, Ky. Advertisement A.S.H.V.E. *Guide*, 1948.

(89) "Study of Changes in Temperature and Water Vapor Content of Respired Air in Nasal Cavity" by L. E. Leeley. A.S.H.V.E., Trans., 1940.

(90) "Entrainment and Jet-Pump Action of Air Streams" by G. L. Tune, G. B. Priester, and D. K. Wright, Jr. A.S.H.V.E., Trans., 1942.

(91) "Control of Air-Streams in Large Spaces" by G. L. Tune and G. B.

Priester. *Heating, Piping and Air Conditioning,* A.S.H.V.E. Journal Section, Jan. 1944.
(92) "A New Development in Absorption Refrigeration" by A. A. Berestneff. *Refrigerating Engineering,* June 1949.
(93) "Carrene 7 A New Refrigerant" by C. M. Ashley. *Refrigerating Engineering,* June 1950.
(94) *Refrigeration and Air Conditioning Directory,* 1952 Edition. Published by Business News Publishing Company, Detroit.
(95) Membership list of Industrial Unit Heater Association, 1952.

BOOKS AND PUBLISHED ARTICLES

BOOKS

Fan Engineering: 1st Edition, 1914, Willis H. Carrier, Editor; 2nd Edition, 1925, R. D. Madison, Editor, Assisted by Willis H. Carrier; 3rd Edition, 1933, R. D. Madison, Editor, Assisted by Willis H. Carrier; Published by Buffalo Forge Company, Buffalo, N.Y.

Modern Air Conditioning, Heating and Ventilating: 1st Edition, 1940, Willis H. Carrier, R. E. Cherne, and W. A. Grant; 2nd Edition, 1950, Willis H. Carrier, R. E. Cherne, and W. A. Grant; Published by Pitman Publishing Corporation, New York, N.Y.

PUBLISHED ARTICLES

1903 "Heating and Ventilating of Foundries and Machine Shops" by Willis H. Carrier. *American Foundrymen's Association Journal,* June 1903; *Foundry* (Abstracted), Vol. 22, 1903; *American Machinist,* August 1903; *Iron Age,* June 1903.
1904 "Cupola Fan Practice" by Willis H. Carrier; American Foundrymen's Association, Trans., Vol. 13, 1904; *Foundry,* October 1904.
1905 "Air Blast for the Foundry Cupola" by Willis H. Carrier; *Iron Age,* May 1905.
1906 "Fan System for Humidifying, Ventilating and Heating Mills" by Willis H. Carrier; *Textile World Record,* April 1906.
1907 "New Departure in Cooling and Humidifying Textile Mills" by Willis H. Carrier; *Textile World Record,* May 1907; *Heating and Ventilating Magazine,* June, July 1907; *Engineering Review* (Abstracted), July 1907; American Cotton Manufacturers Association, Trans., 1907.
1908 "Positive and Accurate Humidity Control" by J. I. Lyle and Willis H. Carrier; National Association of Cotton Manufacturers, Trans., 1908.

1910 "Air-Conditioning Apparatus, Its Construction and Application" by Willis H. Carrier; *Engineers' Society of Western Pennsylvania,* July 1910; *Heating and Ventilating Magazine* (Abstracted), September 1910.

1910 "Notes on the Theory and Present Practice of Humidifying" by Willis H. Carrier; *Heating and Ventilating Magazine,* September 1910.

1911 "Rational Psychrometric Formulae" by Willis H. Carrier; American Society of Mechanical Engineers, Trans., 1911; *A.S.M.E. Journal,* November 1911; *Heating and Ventilating Magazine* (Abstracted), Dec. 1911; *Engineering News,* Vol. 66, 1911; *Industrial Engineering,* Vol. 11, 1911; *Institution of Civil Engineers,* Vol. 188, 1912.

1911 "Air-Conditioning Apparatus" by Willis H. Carrier and F. L. Busey; American Society of Mechanical Engineers, Trans., Vol. 33, 1911; *A.S.M.E. Journal,* December 1911; *Heating and Ventilating Magazine* (Abstracted), March 1912; *Engineering News,* Vol. 66, 1911; *Engineering Review,* Jan., April, May, Sept. 1912; *Institution of Civil Engineers,* Vol. 188, 1912; *Industrial Engineering,* Vol. 11, 1912.

1912 "Rate of Heat Transmission in Indirect Surface Air Heaters" by Willis H. Carrier and F. L. Busey; *Heating and Ventilating Magazine,* January 1912.

1912 "Elements of Design of Humidifiers and of Air-Conditioning Apparatus" by Willis H. Carrier and F. L. Busey; *Heating and Ventilating Magazine,* March 1912.

1913 "Design of Indirect Heating Systems with respect to Maximum Economy of Operation" by F. L. Busey and Willis H. Carrier; American Society of Heating and Ventilating Engineers, Trans., 1913; *Heating and Ventilating Magazine* (Abstracted), March 1913.

1913 "Heating and Ventilating of the Foundry" by Willis H. Carrier; American Foundrymen's Association, Trans., 1913; *Heating and Ventilating Magazine* (Abstracted), January 1913.

1917 Discussions by Willis H. Carrier of Engineering Papers on "Elimination of Corrosion in Hot Water Supply Pipe," "Comparison of Pipe Coils and Cast Iron Sections for Warming Air," "Artificial Drying with the Use of Gas," and "Papers on Drying"; A.S.H.V.E., Trans., 1917.

1918 "Temperature of Evaporation" by Willis H. Carrier; American Society of Heating and Ventilating Engineers, Trans., 1918.

1918 Discussions by Willis H. Carrier of Engineering Papers on: "Economy in Heating," "Fuel Economy in House Heating," "Reasons for Failure of Heating Systems," "High Temperature Drying," and "What We Do and Don't Know About Heating"; A.S.H.V.E., Trans., 1918.

163

1919 Discussions by Willis H. Carrier of Engineering Papers on: "Arrangement of Heating Pipes, an Important Factor in Decay of Factory Roofs," and "Mine Ventilation"; A.S.H.V.E., Trans., 1919.

1920 Discussion of Papers on "Drying" by Willis H. Carrier; A.S.H.V.E., Trans., 1920.

1921 "Compartment Dryer" by Willis H. Carrier and A. E. Stacey, Jr.; *Journal of Industrial and Engineering Chemistry,* May 1921.

1921 "Theory of Atmospheric Evaporation—with Special Reference to Compartment Dryers" by Willis H. Carrier; *Journal of Industrial and Engineering Chemistry,* May 1921.

1921 Discussion of "Theory of Dust Action" by Willis H. Carrier; A.S.H.V.E., Trans., 1921.

1922 "Manufactured Weather and Personal Efficiency" by Willis H. Carrier; *Chemical and Metallurgical Engineering,* August 1922.

1922 Discussions by Willis H. Carrier of Engineering Papers on: "Heat Transfer by Conduction and Convection," and "The Intermediate or Junior High School in Detroit"; A.S.H.V.E., Trans., 1922.

1923 Discussions by Willis H. Carrier of Engineering Papers on: "Modern Tendencies of Ventilation Practice," "Lines of Equal Comfort," "Heat Transmission Through Building Structures," "Heating the New Navy Gun Shop," and "Capacities of Steam Heating Risers as Affected by Critical Velocity of Steam and Condensate Mixtures"; A.S.H.V.E., Trans., 1923.

1924 "Thermodynamic Characteristics of Various Refrigerating Fluids" by Willis H. Carrier and R. W. Waterfill; *Refrigerating Engineering,* June 1924.

1924 "Temperature of Evaporation of Water into Air" by Willis H. Carrier and D. C. Lindsay; American Society of Mechanical Engineers, Trans., 1924; *Mechanical Engineering* (Abstracted), May 1925; *Refrigerating Engineering,* Jan. 1925.

1924 Discussions by Willis H. Carrier of Engineering Papers on: "The Efficiency of Air Cleaners," "Measuring Heat Transmission in Building Structures," "Cooling Effect at Various Air Velocities," and "Ventilation of Department Store"; A.S.H.V.E., Trans., 1924.

1925 Discussions by Willis H. Carrier of Engineering Papers on: "Modern Steam Power Station," and "A Proposed Method for Comparison of Effectiveness of Indirect Heating Surfaces"; A.S.H.V.E., Trans., 1925.

1926 "Centrifugal Compression as Applied to Refrigeration" by Willis H. Carrier; *Refrigerating Engineering,* Feb. 1926.

1926 Discussions by Willis H. Carrier of Engineering Papers on: "Heating Effect of Radiators," "Proposed Code for Rating—Testing Radiators," and "New Psychrometric Humidity Chart"; A.S.H.V.E., Trans., 1926.

1927 Discussion on "How Design and Operation of Heating Plant Compare in an Insulated Office Building" by Willis H. Carrier; A.S.H.V.E., Trans., 1927.

1927 "Thermal Engineer" by Willis H. Carrier; *Refrigerating Engineering*, January 1928.

1928 "Air Conditioning" by Willis H. Carrier; *Encyclopedia Britannica*, Fourteenth Edition, Vol. 1, 1929.

1928 "Air Conditioning for Homes in 1950" by Willis H. Carrier; *Domestic Engineering*, May 1928.

1928 Discussions by Willis H. Carrier of Engineering Papers on: "A Study of Dust Determinators," "Heat Transmission Research," "Method of Calculating Cost of Gas Heating," "Refrigeration as Applied to Air Conditioning," "Standard Test Code for Heat Transmission through Walls," and "Test Methods for Radiators and Boilers"; A.S.H.V.E., Trans., 1928.

1929 "Air Conditioning—Its Phenomenal Development" by Willis H. Carrier; *Heating and Ventilating*, June 1929.

1929 "Air Conditioning in the Printing and Lithographing Industry" by Willis H. Carrier and R. T. Williams; American Society of Mechanical Engineers, Trans., Vol. 51, 1929. *Zeitschrift für die Gesamte Kälte—Industrie* (Abstracted in German), Sept. 1930.

1929 "New Prospects for an Established Industry" by Willis H. Carrier; *Heating, Piping and Air Conditioning*, May 1929.

1929 "Control of Humidity and Temperature as Applied to Manufacturing Processes and Human Comfort" by Willis H. Carrier and others; World Engineering Congress, Tokyo, 1929. Proceedings 1931; *Heating and Ventilating* (Abstracted), Nov., Dec. 1929; (Revised 1939) *Heating, Piping and Air Conditioning;* (Revised 1950) *Heating, Piping and Air Conditioning*, April, June, August 1950, Feb. 1951.

1929 Discussions by Willis H. Carrier of Engineering Papers on: "Architectural Aspects of Concealed Heaters," "Capacity of Radiator Supply Branches for One- and Two-Pipe Systems," "Frost and Condensation on Windows," "Heating with Steam Below Atmospheric Pressure," "Instruments for the Measurement of Air Velocity," "Report of Committee on Rating Low-Pressure Heating Boilers," "The Summer Comfort Zone; Climate and Clothing," and "Thermal Resistance of Air Spaces"; A.S.H.V.E., Trans., 1929.

1930 "Control of Humidity and Temperature" by Willis H. Carrier; *Heating and Ventilating*, March 1930; *Domestic Engineering*, June, July 1930.

1930 "Air Conditioning for Summer and Winter Comfort" by Willis H. Carrier; *Heating, Piping and Air Conditioning*, Sept. 1930; *Plumbers Trade Journal*, Oct. 1930; *Domestic Engineering*, Oct. 1930.

1930 Discussion by Willis H. Carrier of Engineering Papers on: "Constancy of Wet-Bulb Temperature and Total Heat Content during Adiabatic Saturation of Dry Air with Water Vapor"; *Refrigerating Engineering*, Vol. 19, 1930.

1930 Discussions by Willis H. Carrier of Engineering Papers on: "Air Conditioning the Halls of Congress," "How Comfort is Affected by Surface Temperatures and Insulation," "Panel Warming," "Rating of Heating Boilers by Their Physical Characteristics," "Report of Continuing Committee on Codes for Testing and Rating Steam Heating Solid Fuel Boilers," and "Wall Surface Temperatures"; A.S.H.V.E., Trans., 1930.

1931 "Cost of Air Conditioning in Department Stores" by Willis H. Carrier; *Electrical World*, Feb. 1931.

1931 "Field of Refrigeration in Air Conditioning and Methods of its Application" by Willis H. Carrier; *Ice and Refrigeration*, March 1931; *Refrigerating Engineering* (Abstracted), March 1931; *Power*, March 1931.

1931 "Making Weather to Order in the Home" by Willis H. Carrier and Margaret Ingels; *Good Housekeeping*, March 1931.

1931 "Recent Progress in Air Conditioning" by Willis H. Carrier; *Refrigerating Engineering*, March 1931.

1931 "Relation of Standardization to Air Conditioning Printing Industry" by Willis H. Carrier; Printing Institute Conference, Washington, D.C., March 1931; *Heating, Piping and Air Conditioning*, April 1931.

1931 "Making Weather for Human Comfort" by Willis H. Carrier; *Domestic Engineering*, July 1931.

1931 "Standards of Heating and Ventilation" by Willis H. Carrier; *School Executives Magazine*, Oct. 1931.

1931 Discussions by Willis H. Carrier of Engineering Papers on: "Heat Emission from the Surfaces of Cast Iron and Copper Cylinders Heated with Low Pressure Steam," "Heating Effects of Radiators," "Air Conditioning in the Bakery," "Air Conditioning for Railway Passenger Cars," "Air Pollution from the Engineer's Standpoint," "Friction Heads in One-Inch Standard Cast-Iron Tees," and "Essential Elements for Determining Heating Plant Requirements"; A.S.H.V.E., Trans., 1931.

1932 "Accomplishments in the Art of Heating and Ventilating, and Contributions Thereto during My Presidential Year" by Willis H. Carrier, Jan. 1932; A.S.H.V.E., Trans., 1932.

1932 "How to Use the Effective Temperature Index and Comfort Charts" by C. P. Yaglon, Willis H. Carrier, E. V. Hill, F. C. Houghten, and J. H. Walker; *Heating, Piping and Air Conditioning*, June 1932.

1932 "Weather Yields to Man's Control" by Willis H. Carrier; *National Safety News,* June 1932.

1932 "Gas and Air Conditioning" by Willis H. Carrier; *American Gas Journal,* October 1932.

1932 "Air Conditioning" by Willis H. Carrier; *Manufacturers Record,* Dec. 1932.

1932 "Air Conditioning Is Saving Industry $15,000,000 Yearly" by Willis H. Carrier; *Heating, Piping and Air Conditioning,* Dec. 1932; *Refrigeration* (Abstracted), Feb. 1933.

1933 "History and Development of Air Conditioning Industry" by Willis H. Carrier; *Refrigeration,* Jan. 1933.

1933 "Trends in Heating and Air Conditioning" by Willis H. Carrier; *Heating, Piping and Air Conditioning,* Jan. 1933.

1933 "Future of Heating and Air Conditioning" by Willis H. Carrier; *Domestic Engineering,* Feb. 1933.

1933 "Present Trends in Air Conditioning Summarized" by Willis H. Carrier; *School Management,* March 1933.

1933 "Economics of Man-Made Weather" by Willis H. Carrier; *Scientific American,* April 1933.

1934 Discussions by Willis H. Carrier of Engineering Papers on: "Study of Fuel Burning Rates and Power Requirements of Oil Burners in Relation to Excess Air," and "Design and Valuation of Cast Iron Domestic Heating Boilers"; A.S.H.V.E., Trans., 1934.

1935 Discussion by Willis H. Carrier on "A Rational Heat Gain Method for the Determination of Air Conditioning Cooling Loads"; A.S.H.V.E., Trans., 1935.

1936 "Progress in Air Conditioning in the Last Quarter Century" by Willis H. Carrier; *Heating, Piping and Air Conditioning,* August 1936; A.S.H.V.E., Trans., 1936.

1936 Discussions by Willis H. Carrier of Engineering Paper on: "Refrigerants—Their Use and Regulation" by H. D. Edwards; *Refrigerating Engineering,* Aug. 1936.

1936 "Air Conditioning—Advances and Trends of the Art" by Willis H. Carrier and R. W. Waterfill; American Institute of Refrigeration, Holland, 1936; Proceedings—International Congress of Refrigeration, 1936.

1936 "Appraising Air Conditioning and Cooling Equipment" by Willis H. Carrier; *Review* (Society of Residential Appraisers), Nov. 1936.

1936 "Air Conditioning" by Willis H. Carrier; *The Encyclopedia Americana,* 1936.

1937 "Review of Existing Psychrometric Data in Relation to Practical Engineering Problems" by Willis H. Carrier and C. O. Mackey; A.S.M.E., Trans., 1937.

1937 "Contact-Mixture Analogy Applied to Heat Transfer with Mixtures of Air and Water Vapor" by Willis H. Carrier; A.S.M.E., Trans., 1937.

1937 "A Pioneer Looks into the Future of Air Conditioning" by Willis H. Carrier; *Automatic Heat and Air Conditioning*, April 1937.

1937 "Applying Science and Engineering to Education" by Willis H. Carrier; *Mechanical Engineering*, May 1937.

1937 "Principles versus Curriculum in Memorizing Facts and Theories" by Willis H. Carrier; *Mechanical Engineering*, May 1937.

1937 "The Graduate's Dilemma: Small City versus Large City" by Willis H. Carrier; *Rotarian*, June 1937.

1938 Discussion by Willis H. Carrier of Engineering Paper on: "Mining Methods and Health and Safety in Mines"; *American Institute of Mining and Metallurgical Engineers*, Feb. 1938.

1938 "The Development of Air Conditioning" by Willis H. Carrier; *Refrigeration and Air Conditioning*, Quebec, Canada, Summer 1938.

1938 "Air Conditioning in Relation to Industrial Hygiene" by Willis H. Carrier; *Industrial Physicians and Surgeons*, Chicago, June 1939; *Industrial Medicine*, September 1938.

1938 "Air Cooling in Gold Mine on Rand" by Willis H. Carrier; *American Institute of Mining and Metallurgical Engineers*, Vol. 141, 1940; *Canadian Mining Journal*, December 1938.

1938 "Weather by Prescription" by Willis H. Carrier; *National Safety News*, October 1938.

1938 Discussions by Willis H. Carrier of Engineering Papers on: "Performance Tests of Asbestos Insulating Air Duct," and "Physiological Reactions and Sensations of Pleasantness under Varying Atmospheric Conditions"; A.S.H.V.E., Trans., 1938.

1939 "Development of Air Conditioning" by Willis H. Carrier; *Western Society of Engineers Journal*, February 1939.

1939 "Making Weather around the Globe" by Willis H. Carrier; *Popular Mechanics*, July 1939.

1939 Discussion by Willis H. Carrier of Engineering Paper on: "Air Conditioning in Industry"; A.S.H.V.E., Trans., 1939.

1940 "Employer Suggests Needed Improvements in our System of Technical Education" by Willis H. Carrier; *Mechanical Engineering*, October 1940; *Electrical Engineering* (Abstracted), September 1940; *Engineer* (and Editorial Comment), November 1940; *Engineering*, January 1941.

1940 "Air Conditioning—An Industry Opportunity" by Willis H. Carrier; *American Gas Association*, 1940; New York *Times* (Abstracted), October 1940.

1941 "Conduit System of Central Station Air Conditioning" by Willis H. Carrier and G. W. Meek; *Refrigerating Engineering*, July 1941.

168

1941 Discussion by Willis H. Carrier of Engineering Paper on: "Local Cooling of Workers in Hot Industry"; A.S.H.V.E., Trans., 1941.

1942 "After the War, What?" by Willis H. Carrier; *Refrigeration Engineering*, September 1942.

1943 "How Air Conditioning Has Advanced Refrigeration" by Willis H. Carrier; *Mechanical Engineering*, May 1943; *Modern Refrigeration*, July 1943; *Machinery Market* (Abstracted), July 1943.

1943 "Trends in Technical Education" by Willis H. Carrier; *Ceramic Age*, May 1943.

1943 Discussions by Willis H. Carrier of Engineering Papers on: "Final Values of the Interaction Constant for Moist Air," "Spray Nozzle Performance in a Cooling Tower," and "Study of Actual vs. Predicted Cooling Load on an Air-Conditioning System"; A.S.H.V.E., Trans., 1943.

1944 "Prospects in Air-Conditioning and Refrigeration Fields" by Willis H. Carrier; *Machinery*, January 1944.

1944 "The Resistance to Heat Flow through Finned Tubing" by Willis H. Carrier and S. W. Anderson; *Heating, Piping and Air Conditioning*, May 1944.

1944 "Air Conditioning for Peace-time Production" by Willis H. Carrier; *The Underwear & Hosiery Review*, April 1944.

1944 Discussions by Willis H. Carrier of Engineering Papers on: "Aerodynamic Development of Axial Flow Fans," "The Discoloration Methods of Rating Air Filters," "A Method of Heating a Corrugated-Iron Coal Preparation Plant," and "A Study of Intermittent Heating of Churches"; A.S.H.V.E., Trans., 1944.

1945 "Air Conditioning Industry" by Willis H. Carrier; *Army and Navy Journal*, September 1945.

1945 "Air Conditioning" by Willis H. Carrier; *Encyclopedia Americana*, 1945.

1945 Discussion by Willis H. Carrier on Engineering Paper: "Summer Weather Data and Sol-Air Temperature"; A.S.H.V.E., Trans., 1945.

1946 "American Developments in Air Conditioning and Refrigeration during War Period" by Willis H. Carrier; *Engineers' Digest* (British Edition), January 1946.

1946 "Let's Hit a High Standard of Living for the American People" by Willis H. Carrier; *Home Comfort Wholesaler*, January 1946.

1946 "Modern Marine Refrigeration and Air Conditioning" by Willis H. Carrier and L. E. Starr; *Refrigerating Engineering*, Vol. 51, 1946; *Marine Engineering and Shipping Review* (Abstracted), April 1946; *Pacific Marine Review*, July 1946.

1946 Discussions by Willis H. Carrier of Engineering Papers on: "Characteristics of Unit Dust Collectors," "Heat Losses through Wetted Walls," "Hold-Fire Controls for Bituminous Coal Stokers," and

"Influence of an Interior Coating of Aluminum Paper on Internal Thermal Conditions and Heat Economy"; A.S.H.V.E., Trans., 1946.

1947 "American Developments in Air Conditioning and Refrigeration during the Last Decade" by Willis H. Carrier; *Institution of Mechanical Engineers*, Vol. 157, 1947.

1947 "Air Conditioning" by Willis H. Carrier; *Lionel Marks Handbook*, 1947.

1951 "Principles of Air Conditioning" by Willis H. Carrier; *Heating, Piping and Air Conditioning*, April, June, August 1950, February 1951.

1953 "Mixing and Equilibrium in Pipe Flow" by Willis H. Carrier and W. S. Misener.
(Scheduled for Spring Meeting)—A.S.M.E., 1953.

TECHNOLOGY AND SOCIETY

An Arno Press Collection

Ardrey, R[obert] L. **American Agricultural Implements.** In two parts. 1894

Arnold, Horace Lucien and Fay Leone Faurote. **Ford Methods and the Ford Shops.** 1915

Baron, Stanley [Wade]. **Brewed in America:** A History of Beer and Ale in the United States. 1962

Bathe, Greville and Dorothy. **Oliver Evans:** A Chronicle of Early American Engineering. 1935

Bendure, Zelma and Gladys Pfeiffer. **America's Fabrics:** Origin and History, Manufacture, Characteristics and Uses. 1946

Bichowsky, F. Russell. **Industrial Research.** 1942

Bigelow, Jacob. **The Useful Arts:** Considered in Connexion with the Applications of Science. 1840. Two volumes in one

Birkmire, William H. **Skeleton Construction in Buildings.** 1894

Boyd, T[homas] A[lvin]. **Professional Amateur:** The Biography of Charles Franklin Kettering. 1957

Bright, Arthur A[aron], Jr. **The Electric-Lamp Industry:** Technological Change and Economic Development from 1800 to 1947. 1949

Bruce, Alfred and Harold Sandbank. **The History of Prefabrication.** 1943

Carr, Charles C[arl]. **Alcoa, An American Enterprise.** 1952

Cooley, Mortimer E. **Scientific Blacksmith.** 1947

Davis, Charles Thomas. **The Manufacture of Paper.** 1886

Deane, Samuel. **The New-England Farmer,** or Georgical Dictionary. 1822

Dyer, Henry. **The Evolution of Industry.** 1895

Epstein, Ralph C. **The Automobile Industry:** Its Economic and Commercial Development. 1928

Ericsson, Henry. **Sixty Years a Builder:** The Autobiography of Henry Ericsson. 1942

Evans, Oliver. **The Young Mill-Wright and Miller's Guide.** 1850

Ewbank, Thomas. **A Descriptive and Historical Account of Hydraulic and Other Machines for Raising Water,** Ancient and Modern. 1842

Field, Henry M. **The Story of the Atlantic Telegraph.** 1893

Fleming, A. P. M. **Industrial Research in the United States of America.** 1917

Van Gelder, Arthur Pine and Hugo Schlatter. **History of the Explosives Industry in America.** 1927

Hall, Courtney Robert. **History of American Industrial Science.** 1954

Hungerford, Edward. **The Story of Public Utilities.** 1928

Hungerford, Edward. **The Story of the Baltimore and Ohio Railroad, 1827-1927.** 1928

Husband, Joseph. **The Story of the Pullman Car.** 1917

Ingels, Margaret. **Willis Haviland Carrier, Father of Air Conditioning.** 1952

Kingsbury, J[ohn] E. **The Telephone and Telephone Exchanges:** Their Invention and Development. 1915

Labatut, Jean and Wheaton J. Lane, eds. **Highways in Our National Life:** A Symposium. 1950

Lathrop, William G[ilbert]. **The Brass Industry in the United States.** 1926

Lesley, Robert W., John B. Lober and George S. Bartlett. **History of the Portland Cement Industry in the United States.** 1924

Marcosson, Isaac F. **Wherever Men Trade:** The Romance of the Cash Register. 1945

Miles, Henry A[dolphus]. **Lowell, As It Was, and As It Is.** 1845

Morison, George S. **The New Epoch:** As Developed by the Manufacture of Power. 1903

Olmsted, Denison. **Memoir of Eli Whitney, Esq.** 1846

Passer, Harold C. **The Electrical Manufacturers, 1875-1900.** 1953

Prescott, George B[artlett]. **Bell's Electric Speaking Telephone.** 1884

Prout, Henry G. **A Life of George Westinghouse.** 1921

Randall, Frank A. **History of the Development of Building Construction in Chicago.** 1949

Riley, John J. **A History of the American Soft Drink Industry:** Bottled Carbonated Beverages, 1807-1957. 1958

Salem, F[rederick] W[illiam]. **Beer, Its History and Its Economic Value as a National Beverage.** 1880

Smith, Edgar F. **Chemistry in America.** 1914

Steinman, D[avid] B[arnard]. **The Builders of the Bridge:** The Story of John Roebling and His Son. 1950

Taylor, F[rank] Sherwood. **A History of Industrial Chemistry.** 1957

Technological Trends and National Policy, Including the Social Implications of New Inventions. Report of the Subcommittee on Technology to the National Resources Committee. 1937

Thompson, John S. **History of Composing Machines.** 1904

Thompson, Robert Luther. **Wiring a Continent:** The History of the Telegraph Industry in the United States, 1832-1866. 1947

Tilley, Nannie May. **The Bright-Tobacco Industry, 1860-1929.** 1948

Tooker, Elva. **Nathan Trotter:** Philadelphia Merchant, 1787-1853. 1955

Turck, J. A. V. **Origin of Modern Calculating Machines.** 1921

Tyler, David Budlong. **Steam Conquers the Atlantic.** 1939

Wheeler, Gervase. **Homes for the People,** In Suburb and Country. 1855